Environmental Software Systems

IFIP – The International Federation for Information Processing

IFIP was founded in 1960 under the auspices of UNESCO, following the First World Computer Congress held in Paris the previous year. An umbrella organization for societies working in information processing, IFIP's aim is two-fold: to support information processing within its member countries and to encourage technology transfer to developing nations. As its mission statement clearly states,

> IFIP's mission is to be the leading, truly international, apolitical organization which encourages and assists in the development, exploitation and application of information technology for the benefit of all people.

IFIP is a non-profitmaking organization, run almost solely by 2500 volunteers. It operates through a number of technical committees, which organize events and publications. IFIP's events range from an international congress to local seminars, but the most important are:

- the IFIP World Computer Congress, held every second year;
- open conferences;
- working conferences.

The flagship event is the IFIP World Computer Congress, at which both invited and contributed papers are presented. Contributed papers are rigorously refereed and the rejection rate is high.

As with the Congress, participation in the open conferences is open to all and papers may be invited or submitted. Again, submitted papers are stringently refereed.

The working conferences are structured differently. They are usually run by a working group and attendance is small and by invitation only. Their purpose is to create an atmosphere conducive to innovation and development. Refereeing is less rigorous and papers are subjected to extensive group discussion.

Publications arising from IFIP events vary. The papers presented at the IFIP World Computer Congress and at open conferences are published as conference proceedings, while the results of the working conferences are often published as collections of selected and edited papers.

Any national society whose primary activity is in information may apply to become a full member of IFIP, although full membership is restricted to one society per country. Full members are entitled to vote at the annual General Assembly, National societies preferring a less committed involvement may apply for associate or corresponding membership. Associate members enjoy the same benefits as full members, but without voting rights. Corresponding members are not represented in IFIP bodies. Affiliated membership is open to non-national societies, and individual and honorary membership schemes are also offered.

Environmental Software Systems

Proceedings of the International Symposium on Environmental Software Systems, 1995

Edited by

Ralf Denzer
Saarland State University for Technology and Business
Germany

Gerald Schimak
Austrian Research Centre Seibersdorf
Austria

and

David Russell
Penn State Great Valley
USA

Published by Chapman and Hall on behalf of the
International Federation for Information Processing (IFIP)

CHAPMAN & HALL
London · Glasgow · Weinheim · New York · Tokyo · Melbourne · Madras

Published by Chapman & Hall, 2–6 Boundary Row, London SE1 8HN, UK

Chapman & Hall, 2–6 Boundary Row, London SE1 8HN, UK

Blackie Academic & Professional, Wester Cleddens Road, Bishopbriggs, Glasgow G64 2NZ, UK

Chapman & Hall GmbH, Pappelallee 3, 69469 Weinheim, Germany

Chapman & Hall USA, 115 Fifth Avenue, New York, NY 10003, USA

Chapman & Hall Japan, ITP-Japan, Kyowa Building, 3F, 2-2-1 Hirakawacho, Chiyoda-ku, Tokyo 102, Japan

Chapman & Hall Australia, 102 Dodds Street, South Melbourne, Victoria 3205, Australia

Chapman & Hall India, R. Seshadri, 32 Second Main Road, CIT East, Madras 600 035, India

First edition 1996

Printed in Great Britain by TJ Press, Padstow, Cornwall

ISBN 0 412 73730 2

A catalogue record for this book is available from the British Library

∞ Printed on permanent acid-free text paper, manufactured in accordance with ANSI/NISO Z39.48-1992 and ANSI/NISO Z39.48-1984 (Permanence of Paper).

CONTENTS

Preface

Due to increasing practical needs, software support of environmental protection and research tasks is growing in importance and scope. Software systems help to monitor basic data, to maintain and process relevant environmental information, to analyze gathered information and to carry out decision processes, which often have to take into account complex alternatives with various side effects. Therefore software is an important tool for the environmental domain.

When the first software systems in the environmental domain grew - 10 to 15 years ago - users and developers were not really aware of the complexity these systems are carrying with themselves: complexity with respect to entities, tasks and procedures. I guess nobody may have figured out at that time that the environmental domain would ask for solutions which information science would not be able to provide and - in several cases - can not provide until today. Therefore environmental informatics - as we call it today - is also an important domain of computer science itself, because practical solutions need to deal with very complex, interdisciplinary, distributed, integrated, sometimes badly defined, user-centered decision processes. I doubt somebody will state that we are already capable of building such integrated systems for end users for reasonable cost on a broad range.

The development of the first scientific community for environmental informatics started around 1985 in Germany, becoming a technical committee and working group of the German Computer Society in 1987. This community grew rapidly and the yearly national conferences are amongst the largest scientific informatics conferences in Germany. In 1992, working group WG 5.11 of the International Federation of Information Processing was founded (see overview article by Bernd Page).

The International Symposium on Environmental Software Systems 1995 goes back on an invitation by my colleague and friend David Russell, Penn State Great Valley, to come to Penn State and organize a symposium as a starting point to gather the international scientific community. Our goal was to bring together experts from all over the world dealing with environmental software - from the practitioners point of view as well as from the computer scientists point of view. Not knowing how the international reaction would be, we were happy when ISESS 1995 proved to be a great success. The days we spent together at the Penn State Campus at Great Valley, Malvern, PA were dominated by high quality presentations and a good spirit.

There was much discussion about whether environmental informatics should be considered as a discipline or just as an application area. If we look at business informatics, nobody will doubt the notion of a discipline. If we also think of an application area in terms of *objects*, *methods* and *procedures*, then it is clear, that business informatics today is able to handle its objetcs and methods and is very advanced in handling of procedures within companies. Nobody will doubt that business informatics has changed the way people work.

In my opinion, the reason why we can not yet see this in environmental informatics is very simple: our objects are much more complex, our methods are far more complex and we are far away from thinking in terms of procedures. To make it short: we are still at the very beginning. Therefore, the discussion whether environmental informatics is a discipline, is not of great importance. This question will disappear as in comparable disciplines.

I wish to thank several persons and institutions for their support during the whole process. First of all, my thanks go to David Russell and Penn State Great Valley for inviting us, for providing the facilities and the conference office, for the marvelous equipment and social facilities. Thanks also go to my friend Gerald Schimak and the Austrian Research Center Seibersdorf for their support. Finally, I wish to thank Jim Alpigini, Lockheed Martin, for acting as local chair and keeping things together, as well as Lockheed Martin for sponsoring the symposium.

Gaiberg, August 1995

Prof. Dr. Ralf Denzer, Editor

PART ONE

Invited Lectures

1

Environmental Informatics Towards a new Discipline in Applied Computer Science for Environmental Protection and Research

B. Page
University of Hamburg, Department of Informatics
Vogt-Koelln-Str. 30, D-22527 Hamburg, Germany

Abstract

Environmental problem solving is very much an information processing activity handling a wide range of environmental data from various sections and of different structure. In this review paper a typology of environmental processing systems is given and the background, terminology and main issues of a novel discipline of applied Computer Science - more recently named as Environmental Informatics - introduced. A selected Environmental Informatics project is presented dealing with the development of a knowledge-based simulation frame system supporting modeling and simulation of environmental systems. It was employed as prototyping tool for designing a specific table-based modeling system for road traffic emission analysis. Finally the emerging research scene of Environmental Informatics in Germany and on the international scale is discussed.

1 INTRODUCTION: THE ROLE OF ENVIRONMENTAL INFORMATION PROCESSING

The protection of our environment is one of the greatest challenges in our industrialized societies. This challenge is addressing politics, economy as well as technology and research. It is clear that the various problems in environmental protection, environmental planning, research and engineering can be only solved on the ground of a comprehensive and reliable information basis. State and dynamics of the environment are described by biological, physical, chemical, geological, meteorological, or socio-economical data. This data is *time* and *space dependent* and addresses past or current states. The processing of this data and the production of *information* on the environment, on its stress factors and mutual influence mechanisms are fundamental for any kind of environmental planning and preventive measures. Therefore environmental problem solving is mainly an information processing activity handling a wide range of environmental data. Solutions to our environmental problems are strongly dependent on the quality of accessable information sources. Certainly, information is a very critical factor in making decisive political actions and in changing people´s attitudes on the environment.

Meanwhile the application of information technology has become absolutely essential in the environmental domain for providing the required environmental information on the appropriate level of detail, completeness, accuracy and speed. This information on environmental aspects is just as important as basis for decisions on actions in environmental protection as for gaining knowledge in environmental research. It includes descriptions of the current state of the environment, prognoses on future environmental developments as well as assessments on current or future states.

In Figure 1 it is indicated, how human caused environmental burdens and damages could be couteracted with computerized environmental data and information for the public, the economy, the administration and politics by means of feedback effects.

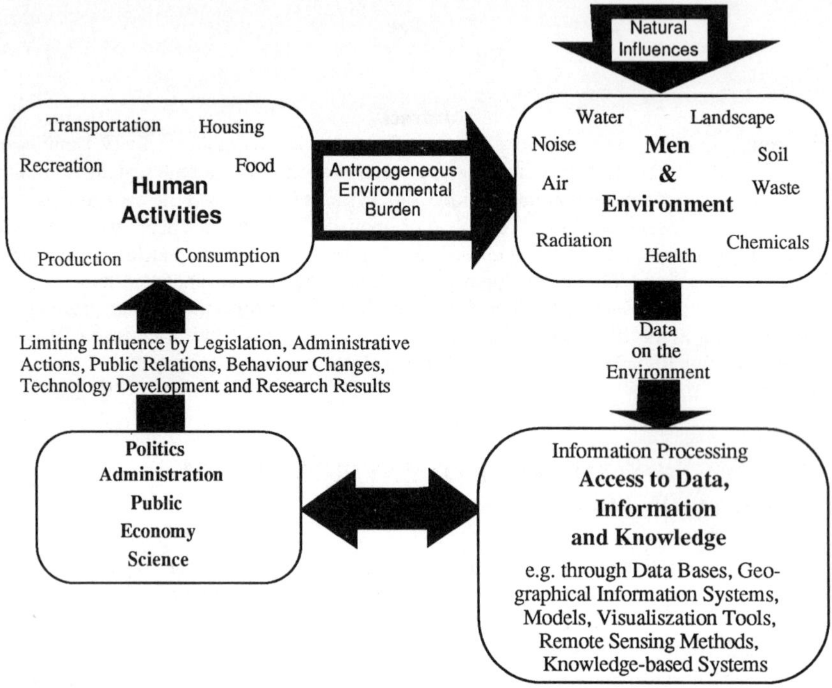

Figure 1 Potential Feedback Effects from Environmental Information Processing.

The following trends in environmental information processing have been recently reported (Radermacher 1994; Hilty 1994a; Hilty 1995a):

- environmental monitoring by means of remote sensing and the combination of data streams from all over the world;
- a policy for sharing and integrating environmental information across political and organizational bounderies;
- advanced model-based data analysis techniques, shifting the focus from data to dynamic system structure;
- industrial applications of environmental information processing, aiming at higher ecological efficiency (resource productivity) of the economic system.

It is obvious that advanced computing technologies play an important role in these developments.

Information technology applications in the environmental domain so far have been often lacking a sound conceptual and scientific basis, since there has not been any significant research in this special domain for a long time. This is certainly not only a matter of applied Computer Science, but an *interdisciplinary task* where many scientific disciplines should be involved (e.g. geo- and bio-sciences, environmental engineering, economics and law, measuring technology, management sciences, etc.). On the other hand, the growing field of environmental information processing is a great challenge to Computer Science methodologies and their applications. From this process of mutual stimulation, a new discipline has been recently emerging, named as *Environmental Informatics*.

2 A TYPOLOGY OF ENVIRONMENTAL INFORMATION PROCESSING SYSTEMS

There is a wide spectrum of environmental information processing systems which can be classified based on the nature of the information and the type of processing. The typology introduced in (Page 1992a) includes monitoring and control systems, conventional information systems, computational evaluation and analysis systems, planning and decision support systems, and integrated environmental information systems:

- *Monitoring and control systems*: Monitoring systems deal with the automation of measurements (including remote sensing) in water, air, soil, noise, and radiation control. This also includes basic data analysis such as the aggregation of time series data, the classification of environmental objects (e.g. in satellite images), or the Identification of hazardous substances based on the measured data. Computerized process control is either directly employed in environmental technology such as in air emission control, sewage, sludge, or refuse processing, or used in production process automation with secondary effects on environmental protection (e.g. energy conservation, emission reduction). Monitoring and control systems often require the processing of vague information.
- *Conventional information systems* are employed for the input, storage, structuring, integration, retrieval, and the presentation of various kinds of environmental information such as raw measurement data, descriptions of environmental objects (such as geographical objects or chemical substances), as well as documents such as environmental regulations or literature references. Spatial and temporal aspects often play an important role in the manage-

ment of these kinds of information. Various types of software tools, including geographical information systems (GIS), or relational data base systems are necessary for coping with these requirements.

- *Computational evaluation and analysis systems* support environmental data processing using complex numerical/statistical analysis methods and modeling techniques. This includes the simulation of various environmental scenarios. Typical applications of these systems are the identification of relevant causes of environmental impacts or the derivation of future effects of different planning measures (e.g. causal models in forest damage research or forecast of emission loads over time and region including possible counter-measures).
- *Planning and decision support systems* help decision makers by offering criteria for the evaluation of alternatives or for justifying decisions, e.g. for environmental impact analyses, for handling hazardous substances, for water resources management, or for technological risk assessment. For industrial applications, computer-based environmental information and management systems are currently being developed, which have to provide the environmental information processing infrastructure for industrial enterprises.
- *Integrated environmental information systems* cannot be uniquely related to the system types mentioned above, because they consist of multiple components serving various purposes. Many complex real world environmental information systems are of an integrated nature (e.g. a state-wide air polution information system with an automated monitoring subsystem, a measurement data base and complex statistical analysis modules with extensive graphical features). It can be expected that integrated environmental information systems will be increasingly designed as distributed systems. The integration of various concepts for information processing, which is required for building these types of systems, is a special challenge to Computer Science methodology.

3 ENVIRONMENTAL INFORMATICS: BACKGROUND, TERMINOLOGY AND MAIN ISSUES

3.1 Scientific Background, Terms and Definitions

The term "Informatics" has been introduced in Europe more than 25 years ago for the *science of the fundamental methodology of information and knowledge processing, its applications as well as its implications* (e.g. on the users, on the work places, on organizations - more recently also on the natural environment as shown in this ISESS conference). Many regard "Informatics" as synonymous to "Computer Science", however, it is argued here that the discipline of "Informatics" has been establishing a broader understanding. Beyond engineering aspects, Informatics is also viewing computer systems as they are embedded in organizations and in society. This is particularily true for *Applied Informatics* (sometimes also called "Applied Computer Science"), the branch of Informatics which bridges the gap between information technology and its various application fields.

Applied Informatics is holding a mediation role (Hilty 1995a). On one hand, it analyses real-world problems in a given application field and defines requirements on information technology. On the other hand, it introduces the problem solving potential of Informatics into the application discipline. In this way the working techniques and eventually the methodologies are

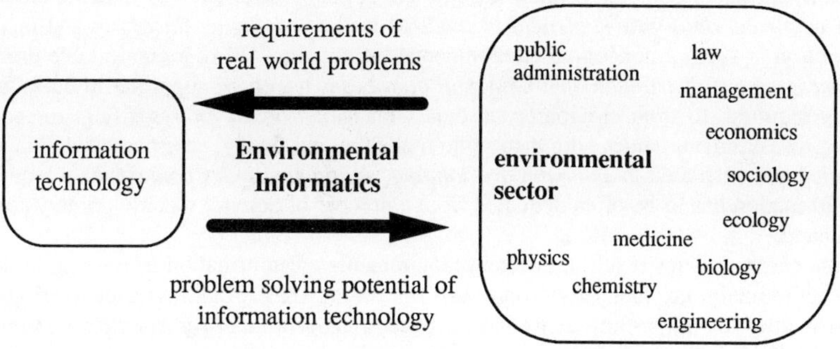

Figure 2 Environmental Informatics supports an multidisciplinary field of Information Technology applications (from Hilty 1995a).

strongly influenced or even changed in many disciplines by an intensified IT utilization. One striking example is medicine where even a completely new speciality called "*Medical Informatics*" was evolving out of medical computer applications as long as 20 years ago. This specialty combines expertise from Applied Informatics with medical knowledge.

A similar situation we find in the environmental sciences today leading to the formation of *Environmental Informatics* (Figure 2). However, we have to point out, that the environmental sector is by itself an interdisciplinary research domain. It incorporates knowledge from biology, ecology, physics, chemistry, engineering, economics, geo-sciences, management, sociology, public administration, law and medicine. This and also the fact that the younger environmental sciences have not yet achieved the level of stability of a classic science like medicine, make it more complicated, but also more challenging for *Environmental Informatics* to play its mediation role. The rather new term *Environmental Informatics* can be defined as follows (Page 1994a and 1994c):

Environmental Informatics is a subdiscipline of Applied Informatics dealing with methods and techniques of Computer Science for analysing, supporting and forming those information processing procedures contributing to the investigation, removal, avoidance and minimization of environmental burden and damages.

3.2 Main Issues

The claim of *Environmental Informatics* as a special subdiscipline of Applied Informatics of its own right can be derived from the specific characteristics of environmental data and their processing requirements typically not found in conventional information processing areas (i.e. business or administrative data processing or even in many engineering computer applications). First, we have many different, rather heterogenous data structures and information sources in environmental information systems such as text data on environmental laws or research projects, measurement data from monitoring networks, structural data on chemical substances, formated engineering data on enviromental technology, just to name a few. In particular, environmental data is often *geographicaly coded,* i.e. information is attached to a particular point or region in space. The represented data objects are often *multidimensional* and have to be described by means of complex geometric objects (e.g. polygons or curves). Second, the processing of *empirical data* with *statistical methods* as well as of *vague*, *uncertain* and *incomplete information* is a major concern in environmental informatics. Third, a comfortable user access on *heterogeneous distributed environmental data bases* has to be supported. Fourth, it is frequently required, to treat environmental data with *model-based analyses* (e.g. to recompute missing measurement values with dispersion models). And finally, environmental data must be presented and evaluated in a *domain overlapping, multidisciplinary context.* As a requirement, this information has to be often deducted from a number of domain specific primary data bases beforehand.

These characteristics result in different requirements on information processing in the environmental domain, i.e. firstly *Environmental Informatics* has to address aspects of geometric data structures and algorithms as well as of modern *Geographical Information Systems* (GIS). Secondly, *Knowledge-based Systems* are of importance in dealing with uncertain environmental knowledge. Thirdly, *modeling software architectures* are useful for statistical analyses as well as for environmental modeling and simulation. A key issue in *Environmental Informatics* is the important *integration* of data, information, and knowledge from various sources in the environmental sector:

- Integrating *data* means overcoming the heterogeneity caused by the variety of operating and database systems, data formats and documentation conventions, program interfaces and software tools, used in different organizations and disciplines.
- Integrating *information* means to enable data to be interpreted meaningfully in different organizational and disciplinary contexts.
- Integrating *knowledge* is a long-term goal aiming at the compatibility of the terminologies and conceptual frameworks of different disciplines and institutions (Hilty 1995a).

Beyond that, there are many links to *heterogeneous data bases* and *distributed information systems*, to *query languages* and *ergonomic user interfaces*, to *scientific visualization* of complex environmental data (see Denzer 1995) as well as to many neighbouring disciplines such as remote sensing (see Günther 1995), measuring technology or business computing.

The need for computer-based environmental management systems in industrial production is a rather new challenge to information technology (see Radermacher 1994). Information systems designed to support those actions an enterprise is taking to manage its effects on the environment are called *industrial environmental information systems*. With the increasing significance of industrial environmental protection as compared to the public sector, where effective environmental information systems and powerful monitoring networks are already operational or under development for some time (Page 1988; Radermacher 1994), *Environmental Informatics* is now looking more closely into design aspects of industrial environmental information systems (Hilty 1995b). This includes the development of computer-based *ecobalances* dealing with the investigation and evaluation of the total mass and energy flow of a given production process in order to assess its environmental impacts (Schmidt 1994).

3.3 Selected Methodologies and Research Topics

In *Environmental Informatics,* a wide range of Informatics methodologies and techniques are basically applied. Here, we want to discuss in brief a few selected Informatics methodologies and research fields which we consider as particularily relevant for environmental information processing. For more details see (Page 1994b and Avouris 1995).

Database and Geographical Information Systems: Database technologies belong without any doubt to the most important tools in environmental information processing (see Page 1995). Many environmental information systems (EIS) can be considered as *extended Geographical Information Systems (GIS)* because of the common spatial reference of most of the data stored. On the other hand, GIS are specialised database systems for spatially structured data (see Bill 1995). Because most environmental data is related to space (i.e. geographical context), GIS are widely used as a basis for environmental information systems. However, EIS also hold thematic data (i.e. environmental facts such as measurement values or chemical substance attributes, environmental documents such as text data on research projects, laws and regulations, or literature) or data with temporal reference, respectively (e.g. land use alterations of restoration areas, or seasonal fluctuations in harmful substance measurements). GIS can neither cope with the problem of handling thematic data adequately nor of managing time series data.

The employment of traditional relational data base concepts for building up EIS is not a satisfactory solution either, because complex environmental objects such as chemical combinations, measurement series, legal documents or surface elements cannot be adequately modeled in form of normalized relations. Two important questions demanding more research in *Environmental Informatics* are, to what extent and how typical EIS applications can be supported by today´s DBMS concepts (i.e. relational, GIS, or information retrieval systems, see Fuhr 1991) and beyond that, which DBMS concepts (e.g. data models for complex objects, or query languages with spatial/temporal predicates) should be provided by future systems in order to have an appropriate data management support to EIS applications (e.g. *object-oriented database systems*, see Cattel 1994; Günther 1993).

Today´s environmental databases do contain a lot of raw data on many environmental aspects, however this data is often not sufficient to derive the information appropriately as requested by the user. Rather, apart from the raw environmental data also *meta-information* (Radermacher 1991) is required. In a broad sense, this includes information about location, time, precision, and revision dates of the data under consideration, as well as descriptions of the data structures and data formats used (Radermacher 1994). Meta-information which is necessary for an adequate interpretation of the environmental data and for user orientation in finding the relevant information sources has been missing in most existing EIS so far. Therefore, environmental meta-information systems are currently developed to overcome the prevailing lack of meta-information in the environmental sector (see Voigt 1995). They are one of the main issues to be addressed in *Environmental Informatics* at this time.

Modeling and Simulation: Modeling and simulation techniques have been employed in the environmental sector for more than three decades. The first applications emerged from water resource management. Today, the following types of simulation models are used for advanced data analysis tasks, for decision support, planning, or for process control:

- Dispersion and quality models for air, water, or soil (e.g. Sydow 1994; Fedra 1994)
- Ecosystem models and models in ecological economics (e.g. Bossel 1994a and1994b),
- Process models as a part of process control systems (e.g. Gilles 1988),
- Models for the prediction of traffic emissions (e.g. Häuslein 1994) and for planning purposes in "eco-logistics" (e.g. Hilty 1994b).

The task of *Environmental Informatics* in environmental modeling is to provide adequate tools that enable users to build simulation models with not too much effort, i.e. by using graphical modeling languages, modeling and simulation program packages or simulation systems, which provide standard modules that can be used as "building blocks" for modeling (see also Häuslein 1994 and section 4 of this paper).

Computer Graphics and Visualization: Employing computer graphics methodology for scientific visualisation of compex environmental data is significant because there is still a lack of knowledge about causal relationships and regularities in environmental systems. In this situation visualization is an important means for transparent data analysis and for the identification of previously unknown structures.

Moreover, graphical presentations are well suited to communicate environmentally relevant facts to decision-makers and to the public (see also Denzer 1995).

Knowledge-based Systems: More recently knowledge-based systems, in particular expert systems or Neuronal Nets (Keller 1995), were also applied to environmental information processing. Knowledge-based approaches are especially relevant for the interpretation of image data from monitoring systems such as aerial photographs and satellite sensor data. They can also be used in supporting environmental modeling and simulation studies (see section 4).

Expert systems have been developed for decision support or planning applications in the following environmental domains (Page 1990; Simon 1995):

- Waste disposal, evaluation of hazardous substances and contaminated sites,
- Environmental impact assessment and environmental planning,
- Application of environmental laws, regulations, and technical instructions,
- Advice in accidents and emergency situations with hazardous chemicals.

However, most of these projects have not yet advanced beyond the prototype state. The utilization of expert system technology in the environmental domain has been limited todate, because

- the spatial and temporal reference of most environmental problems require substancial advances in *spatial* and *temporal reasoning methods,*
- the *realtime capability of expert systems* is not sufficient yet for many environmental monitoring and environmentally adapted process control applications,
- the *integration* of *expert system techniques* with *conventional system components* (i.e. data bases, geographical information systems, simulation models) has to be improved,
- *knowledge aquisition techniques* and tools for *multidisciplinary knowledge domains* such as environmental sciences have to be refined. In this way, domain knowledge from experts of different disciplines has to be aquired and synthezised.

- *model-based reasoning* is more promising for some environmental domains (e.g. technical diagnosis for wastewater treatment plants, or ecosystem analysis) than straigthforward rule-based techniques, found in most conventional ("flat") expert systems (Page 1990).

Systems Integration: As already mentioned in section 3.2, integration is one key issue in *Environmental Informatics*, and it entails more than purely technical problems. The harmonization of environmental information at national, international, or even worldwide levels is of central importance for gaining a reliable reporting system on the state of the environment. These requirements, however, are confronted with the existing heterogeneity of the different standards, terms and measuring methods in the environmental field as well as of hardware and software environments, database systems, method and model bases, network technology, and programming languages. Approaches for overcoming the heterogeneity, one of the major obstacles to open software solutions, can have a tremendous impact on the productivity in environmental management, and can also lead to a greater stability in system design and system usage in environmental information processing (Riekert 1994; Hilty 1995a).

One possible approach to cope with heterogenous system components could be the installation of *meta-information servers* to facilitate access to distributed information systems which has already proven to be very useful for the integration of monitoring data from different sources (Schimak 1994).

4 AN ENVIRONMENTAL INFORMATICS RESEARCH AND DEVELOPMENT PROJECT: A KNOWLEDGE-BASED SIMULATION FRAME SYSTEM FOR PROTOTYPING ENVIRONMENTAL MODELING TOOLS

4.1 Background

Modeling environmental systems (e.g. road traffic pollution) requires powerful software tools which account for the heterogeneous environmental domain with a wide range of modeling methodologies involved and for non-technical user groups with only limited background in modeling and computing.

Road Traffic is known as one of the most critical sources of environmental problems. Frequently rather rough, highly aggregated traffic emission models are developed on the national scale for governmental decision support by the German Federal Environmental Agency (*Umweltbundesamt*) aiming at the estimation of future road traffic development, of future pollutant emissions, and at the analysis of alternative traffic development scenarios (e.g. laws, regulations, technical standards, taxation measures, new engine technologies). The modeling follows a rather straightforward approach based on statistical time series data easily available at the time of governmental request (usually on short notice). Typically the models are "throw away"-models not very suitable for later use after the completion of the current investigation. Because the computations were quite simple, spread sheet programs were used. However, when the models grew larger the limitations of these tools became obvious (e.g. lack of model transparency and extensionabilty).

It was realized that a new, more flexible modeling tool was required for coping with the increased complexity and number of modeling activities for road traffic emission forecasting in the Federal Environmental Agency. Along with the development of a more flexible and transparent modeling tool a generalized, better structured modeling approach was striven for. A selfmade adaptable simulation system was employed in this applied research project to develop a new table-oriented modeling approach for road traffic emissions based on statistical time series data and for prototyping an appropriate modeling tool of higher flexibility and transparency for this specific modeling domain. The project was sponsored by the German Federal Environmental Agency.

4.2 A Knowledge-based Simulation System as Development Tool

A simulation tool named DYNAMIS IIx was developed at the University of Hamburg as a research prototype. It is gaining its flexibility from making use of object-oriented and knowledge-based techniques and was aimed at improving user support and modeling flexibility at the same time (Häuslein 1993). This includes comfortable graphical modeling, hierarchical model design with any number of aggregation levels, provision of simulation specific knowledge, powerful internal model representation (also including non-numerical knowledge), model management facilities, selection from different simulation methodologies in the system, uncomplicated system extensions by additional simulation methods, or manifold model utilization, respectively.

Graphical Modeling: Model structure is presented graphically by a model diagram in a special system diagram window. The diagram is built up by selecting predefined model variable symbols from a menu under mouse control and by linking them through connecting lines in a way consistent with the given simulation methodology. Thereby, a significant part of model design is carried out on the graphical level. The equations for the model variables required for specifying the complete mathematical model definition are edited in a different window - the content window - existing for particular diagram symbols (i.e. model variables) and opened by clicking on the related symbol of the diagram. Numerical values and arithmetic expressions are entered for describing dynamical behavior on the respective model variables. Quite often this information is already inherent in the model structure diagram and can therefore be generated automatically. In this case, it only has to be interactively supplemented by the modeler. The system also allows for the introduction of user defined, application specific graphical notations (e.g. tree or vehicle icons) to account for the needs of non-technical user groups in certain application areas. The redefinition is only addressing the external model representation and has no impact on the internal system level.

Graphical modeling offers a descriptive and intuitive access to model building from the user point of view. Graphical models can be constructed without programming effort and proficiency. Model diagrams are supporting model documentation already in early model building stages and allow for an easier model understanding by outside users.

Hierarchical Modeling: Another important system feature is the support of modular, hierarchical model construction which proved to be very helpful in the environmental field for building comprehensive environmental models out of separate model building blocks. Each model can be embedded in another model as a submodel. A model interface is generated for each submodel where its external relations to the higher level model or other submodels are defined. In this way, a model can be constructed from other (sub) models across any number of aggregation levels. With this approach supported by an appropriate model management facility the utilization of already existing models as model components is promoted.

Knowledge-based Support: Simulation specific knowledge is included in the simulation system making use of knowledge-based techniques to support inexperienced users in model building proficiency. Knowledge bases provided by the system incorporate knowledge on given simulation methodologies (i.e. knowledge on methodologically correct model construction). This knowledge is represented as assertions and rules. It is provided to the user in different forms. User actions are checked for consistency with these rules during the model building process. This includes the model symbol placement and the line connections; e.g. not all model variables are combinable with each other or can be only connected with special links. Rule violation, i.e. an inconsistency with simulation methodology, leads to warnings or even user action breakoffs in severe cases. With the explanation facility of the embedded expert system shell further comments can be generated explaining the error sources in more detail and giving alternative advice. Beyond the consistency check for each model symbol linked into the model diagram during stepwise model building correctness of the overall model is established before the simulation run is actually started. Here model completeness is analysed, i.e. whether all required variables and links have been specified in the appropriate manner. Other points in the checking procedure concern unique naming conventions for model variables as well as the existence of cycles in model links.

In addition, a model specific knowledge base is established for each model where declarations on the model and its components are entered by the system during the model building process. This knowledge can be retrieved by simple queries or used for deriving new knowledge depending on the available rule set. Thus, it will be possible to make earlier use of knowledge already present in the model for inferences well before a complete numerical model definition. In this way, we allow for an incremental model construction by exploiting inherent model knowledge making inferences on the model long before a mathematical solution (e.g. of differential equations) becomes feasible (Häuslein 1991).

System Extensionability: The employment of object-oriented approaches in system architecture and implementation as well as the integration of an expert system shell allow for high flexibility in system utilization and extension as well as for user support in model building. It includes several different simulation methodologies relevant to environmental modeling (e.g. System Dynamics or Energy-Flow-Models) and offers advice on their employment under specified conditions. The integration of additional simulation methodologies into the running system has been one of the main design goals and therefore has been taken care of in the system architecture; i.e. system extensions for new simulation methodologies are realized by using and supplementing a predefined class hierarchy of model objects based on the inheritance principle.

System architecture (see Figure 3) is characterized by three different layers, each incorporating respective functionality of a given type. First, there is an external layer with all functions serving for model user interface generation and in particular for graphical model presentation in diagram form. The internal layer specifies model contents independently from their external representation form, i.e. it takes care of model evaluation and simulation processing. The

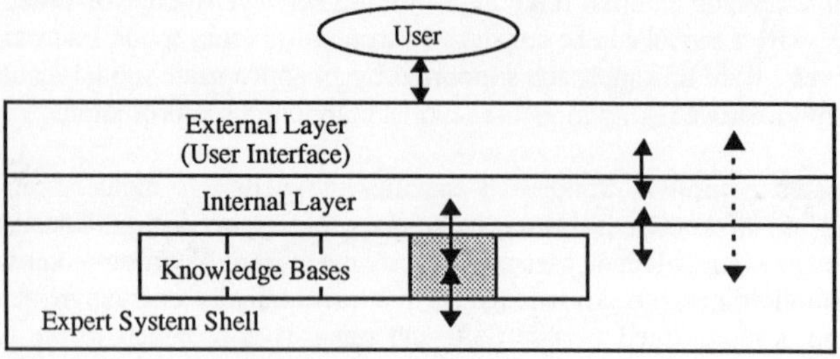

Figure 3 General Architecture of the Knowledge-based Simulation System DYNAMIS IIx

knowledge-based layer incorporates special knowledge on model construction, correctness and classification in several model bases as stated above. Knowledge-based functions are provided by an embedded expert system shell not directly visible on the user interface. The user is working with one homogeneous simulation system from his point of view.

Basic functionality is offered by a system kernel on any layer with all relevant system functions independently of a particular simulation methodology. On each layer the system kernel consists of an object class hierarchy with related methods. In this way, a *simulation frame system* is provided which can be extended for serving concrete simulation methodologies by limited supplements. The extension of the frame system with objects and methods required for supporting a special simulation method can be easily implemented by refering to the available basic object classes and their methods making use of the inheritance mechanism in object-oriented programming (the system was implemented in Allegro Common Lisp with Object Lisp on an Apple Macintosh).

The separation into three system layers facilitates the simulation system extension and modification. Changes can be carried out at the different layers without affecting the other layers. In integrating new simulation methodologies it might well be sufficient to exchange the graphical notation on the external level, introduce a few supplements to the existing object classes on the external level and to substitute the methodological knowledge base. Therefore the simulation system seems well suited for prototying new modeling tools.

4.3 The Prototype Modeling Tool for Road Traffic Emissions

Requirements: The construction of models for road traffic emission analysis can be facilitated by an appropriate software tool meeting a number of basic requirements from this modeling domain. The tool has to provide powerful table processing facilities, since the methodology for road traffic emission on the national scale is basically *table-oriented*, i.e. supporting input of empirical table data, the generation of new tables and the combination of table values. Model building should go hand in hand with an automatic consistency check relating to the modeling methodology, with a clear representation of model structure in graphical form, and should allow for the processing of historical as well as prognosis data, an easy model adaptability and straightforward formula checks. Also there was a requirement for a clear graphical presentation and a structured organization and documentation of simulation results with an interactive selection and output of single model variable dynamics.

The Modeling Approach: Since the data in the field of road traffic emission modeling is typically found in tabular form, the modeling approach to be employed had to take account for the construction and arithmetic manipulation of tables. A new flexible modeling approach named "Dynamic Relations" was introduced combining tabular operations with graphical modeling. The basic concepts for the utilization of tabular data structures ("relations") have been well established in data base theory for a long time already ("Relational Algebra"). We adopted a wide range of these relational operations for the modeling approach, however, we take a different view on relations by adding another dimension - the *time* axis. We supplement the relational operations with some handy functions for arithmetic operations on column values. We introduce interpolation as well as prognosis functions to account for missing or future values, respectively. In this way a simulation run is on one hand reproducing historical progress (e.g. of past emission development) on the basis of empirical time series data (e.g. on vehicle stock, type distribution or mileage) and is on the other hand constructing future scenarios by performing projections and prognoses.

The modeling operations defined in the "Dynamic Relations"-Approach allow for the combination of tables, the arithmetic computation of value sets, the projection and selection of tables as well as the aggregation and grouping of values. The behavior of model variables is specified in definitions, i.e. mathematical expressions evaluating to table constructions. Links in the model diagram describe value flows between model variables. The values of input tables are already available on external files at the start of a simulation, the other model variables generate their specific values by evaluating their definition time step by time step during a simulation run. An overview of the model variables in the "Dynamic Relations"-Approach including the graphical symbols is given in Figure 4. The methodology was implemented as special system extension of the knowledge-based simulation tool DYNAMIS IIx introduced above.

Computational Table	Combination of Tables	
Input Table	Loading of external Data	
Key	Construction of Key Expression	
Selector	Selection and Projection	
Output-Selector	Presentation of a Value Set as Curve	
Operation Instruction	Construction of arithmetic Expression	
Operator	Arithmetic Combination of different Table Columns	
Aggregator	Aggregation of Table Rows	
Submodel	Separation into Submodels (hierarchical Modeling)	
Inout	Communication between Submodels	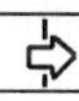

Figure 4 Modeling Components of "Dynamic Relations"-Approach

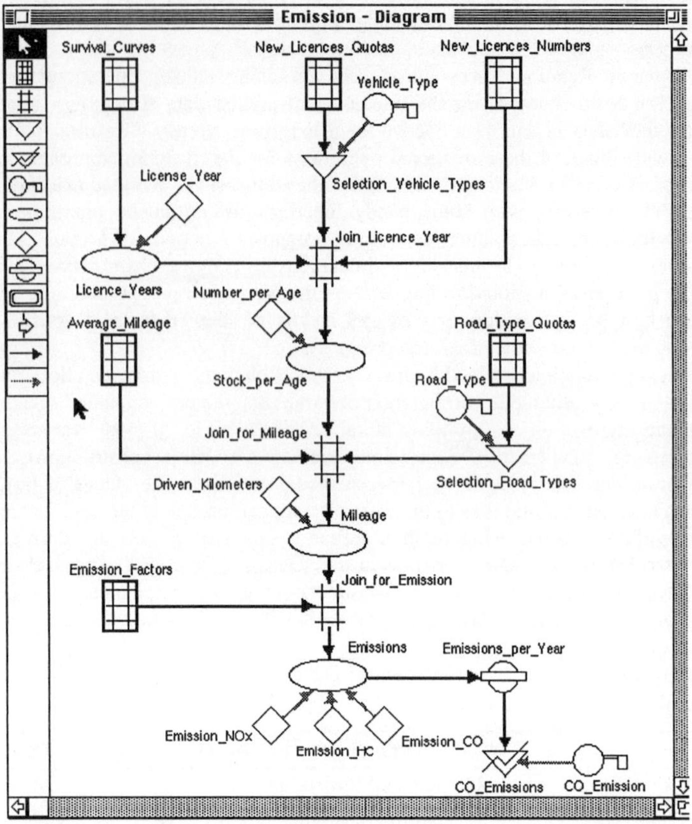

Figure 5 "Dynamic Relations"- Model Diagram for Road Traffic Emission Model

A Road Traffic Emission Model: A rather straightforward emission model for the private car sector in the Federal Republic of Germany for the years 1965 to 2010 was then set up to demonstrate the table-oriented, graphical modeling methodology just introduced (see Figure 5). Input tables include vehicle survival curves, new licence quotas, vehicle type quotas, milage, road types, and emission factors. The dynamic emission computation is carried out in three main steps each producing a new table, i.e. motor vehicle stock (classified by vehicle type and age from survival quotas, licence quotas and overall new licences), mileage per year (for reference year from average mileage classified by vehicle type, age and road type) as well as emissions (per pollutant by multiplication of driven kilometers with emission factors). The model is reproducing or forecasting, respectively the development of the pollutant emissions over the years under consideration. Model results would be produced interactively showing emission curves for different pollutants at the display in the course of the simulation. More model detail could be incorporated without much effort as improved and more differentiated data become eventually available. It would only involve slight changes to introduce refined classifications (i.e. extended tables) or additional types of model variables (e.g. tables with driving cycles).

Assessment: It can be concluded that in the course of the project the research prototype simulation tool with its object-oriented and knowledge-based architecture proved to be a very powerful and flexible prototyping tool for designing special modeling software (Häuslein 1995). From our point of view the "Dynamic Relations"-approach estalished in this project is well suited for table-oriented traffic emission modeling, but its applicability has definitely a much wider scope.

5 THE EVOLVING RESEARCH SCENE IN GERMANY

Scientific Associations and Publications: The new discipline of *Environmental Informatics* has been maturing since the mid-eighties. In Germany, this is well documented by the proceedings of the annual conferences which have been organized by the Special Interest Group "Informatics in Environmental Protection“ in the German Society for Informatics (Gesellschaft für Informatik - GI). This very dedicated group started its activities already back in 1987 and has now around 900 members. A number of working groups have been established dealing with special aspects of environmental information processing such as *Environmental Databases, Visualization of environmental Data, Methods and Tools for environmental Impact Assessment, Integration of environmental Data, Tools for environmental Modeling and Simulation* and more recently *Industrial environmental Information Systems* - just to name some of the working groups.

The anual conferences and a number of additional workshops have yielded a growing number of participants, projects, and publications (e.g., Jaeschke 1987a and b; Valk 1988; Jaeschke 1989; Pillmann 1990; Hälker 1991; Denzer 1992; Jaeschke 1993). Only last year an own book series for the publications of the special interest group has been arranged by a publisher (see Hilty 1994a)

The German Federal Environmental Office is operating an online database "Environmental Literature" and has established an own section "Environmental Data Processing and Informatics" already in 1988. Since then, the relevant German and also selected English publications in this field are registered on a regular basis. An extract of this database section was documented in a bibliography (Umweltbundesamt 1989).

Only recently an own volume "Environmental Informatic up to date" was published in the renowed German Series "Handbook of Informatics" (Page 1994b) and in this way it is documented that our discipline has been accepted now as one of the established Informatics branches. Beyond that, the German Society for Informatics has claimed the issue *"Informatics and the Environment"* as one of their two central themes for the future (together with *"Informatics and Mobility"*; see Jeaschke 1994). Thus, we argue that *Environmental Informatics* is now recognized as integral part of Applied Informatics in Germany.

Education: A few professional schools in Germany are offering continueing education programs in Environmental Information Processing. On the academic level the Department of Informatics at the University of Hamburg has, as the first university in Germany, just started a new elective "Methods of Informatics for Environmental Protection" within the Applied Informatics specialization of the Informatics degree program.

Internationalization: The International Federation for Information Processing (IFIP) has established a working group for this application field (WG 5.11 "Computers and the Environment") in 1992, where members of the German Special Interest Group are playing an active role. The first iniative was a workshop "Informatics for Environmental Protection" at the 12th World Computer Congress in Madrid, Spain in September 1992 (Aiken 1992). In the following year a more specialized working conference was held at Como, Italy dealing with "Computer Support for environmental Impact Assessment" (Guariso 1994) drawing an international participation from more than 15 countries. On the 13th IFIP World Computer Congress in Hamburg, Germany in 1994 a few papers and a panel discussion also focused on environmental information processing aspects (e.g. Radermacher 1994; Hilty 1994c). The new book "Environmental Informatics – Methodology and Applications of Environmental Information Processing" published by Kluwer Academic in English in 1995 (Avouris 1995) - and not to forget this ISESS conference - will eventually promote further the international attention on the new field of *Environmental Informatics*.

6 CONCLUSIONS AND OUTLOOK

Environmental Informatics has now become an integral part of Applied Informatics. Its special characteristic is the transformation of Computer Science methodologies into the application field environmental protection. On one hand, Environmental Informatics has a problem solving potential that must be developed into practical solutions for environmental information processing problems. On the other hand, new and challenging requirements arise in the environmental sector that stimulate research and development in Informatics methodology and applications. The main challenge to Informatics is the complexity and heterogeneity of the environmental sector, demanding for innovative approaches to control complexity and to integrate existing data, information, and knowledge from various scientific disciplines and organizations (Hilty 1995a). It has to be stressed, however, that Environmental Informatics does not only explore the potential benefits of information technology for contributing to the solution of environmental problems, but also sees its responsibility to consider the negative environmental impacts of IT (Rolf 1994).

Also, the role of Environmental Informatics should not be restricted to computer-based investigation of the present state of the environment. Applying sophisticated computing methodologies just to cure at symptoms of our environmental crisis is certainly not enough. Since the Rio Summit has accepted the model "*sustainable development*", meaning a way of living and a form of using resources that does not discriminate against future generations (WCED 1987), there seems to be now a rewarding framework and direction also for our future work at hand. Although it is certainly difficult to operationalize this model, it is clear that providing adequate *information* will be an important step towards changing attitudes and making decisive political actions into that direction possible (Friend 1991). Thus, Environmental Informatics should also accept the challenge to play an active role in the local and global transformations that will be needed to approach sustainability (Hilty 1994c; Rolf 1994).

6 REFERENCES

Aiken, R.M. (eds.) (1992) Education and Society. Information Processing 92. Symposium 2: Informatics for Environmental Protection. IFIP Transactions A-13.Vol. II, pp. 593-686. Amsterdam: North-Holland.

Avouris, N., Page, B. (eds.) (1995) Environmental Informatics – Methodology and Applications of Environmental Information Processing. Kluwer Academic Publishers, Dordrecht, Netherlands.

Bill, R. (1995) Spatial Data Processing in environmental Information Systems, in *Environmental Informatics – Methodology and Applications of Environmental Information Processing* (eds. N. Avouris, B. Page), pp. 53-73.

Bossel, H. (1994a) Modeling and Simulation. A. K. Peters, Wellesley MA.

Bossel, H. (1994b) Understanding Dynamic Systems: Shifting the Focus from Data to Structure, in *Informatik für den Umweltschutz* (eds. L.M. Hilty, A. Jaeschke, B. Page, A. Schwabl), Proc. 8. Symposium, Hamburg 1994, Volume I, pp. 63-75.

Cattel, R.G.G. (1994) The Object Database Standard: ODMG-93. Morgan Kaufmann.

Denzer, R., Schimak, G., Haas, W. (eds.) (1992), Visualisierung von Umweltdaten. Proceedings 3. Workshop, Schloß Zell an der Pram, Dezember 1992. Informatik Aktuell, Berlin, Springer.

Denzer, R., Mayer, H.F., Haas, W. (1995) Visualisation of environmental Data, in *Environmental Informatics – Methodology and Applications of Environmental Information Processing* (eds. N. Avouris, B. Page), pp. 75-92.

Fedra, K. (1994) Model-based Environmental Information and Decision Support Systems, in *Informatik für den Umweltschutz* (eds. L.M. Hilty, A. Jaeschke, B. Page, A. Schwabl), Proc. 8. Symposium, Hamburg 1994, Volume I, pp. 37-58.

Friend, A.M., Rapport, D.J. (1991) Evolution of macro-information systems for sustainable development. *Ecological Economics*, 3/1991, pp. 59-76.

Fuhr, N. (1991) An Information Retrieval View of Environmental Information Systems, in *DEXA 91*, Proc. 2nd Intern.Conf. on Data Base and Expert System Applications, Berlin: Springer.

Gilles, E.D., Marquardt, W. (1988) Prozeßsimulation – ein Beitrag zum aktiven Umweltschutz, in *Beitrag der Mikorelektronik zum Umweltschutz*, Proceedings, GME-Fachbericht. Berlin: vde-Verlag.

Guariso, G., Page, B. (eds.) (1994) Computer Support for Environmental Impact Assessment. IFIP Transactions B-16. Amsterdam: North-Holland.

Günther, O., Riekert, W.-F. (1993) The Design of GODOT: An Object-Oriented Geographic Information System, IEEE Data Engineering Bulletin 16(3), September 1993.

Günther, O., Radermacher, F.J., Riekert, W.-F. (1995) Environmental Monitoring: Models, Methods ans Systems, in *Environmental Informatics – Methodology and Applications of Environmental Information Processing* (eds. N. Avouris, B. Page), pp. 13-38.

Hälker, M., Jaeschke, A. (ed.) (1991) Informatik für den Umweltschutz – Computer Science for Environmental Protection. Proceedings 6. Symposium München, Dezember 1991, Informatik-Fachberichte 296, Berlin: Springer.

Häuslein, A., Page, B. (1991) Knowledge-based Approaches to Modeling and Simulation Support, in *Systems Analysis, Modeling and Simulation 8*, 4/5, (1991), pp. 257-272.

Häuslein, A. (1993) Wissensbasierte Unterstuetzung der Modelbildung und Simulation im Umweltbereich - Konzeption und prototypische Realisierung eines Simulationssystems. Europaeische Hochschulschriften, Series XLI Informatics, Vol. 12, Frankfurt: Peter Lang Publ.
Häuslein, A., Page, B. (1994) Prototyping environmental Modeling Tools with a knowledge-based Simulation Frame System, *in Proc. 14th IMACS World Congress on Computation and Applied Mathematics*. Atlanta, Georgia, July 11-15, 1994, Vol. 3, Atlanta: Georgia Institute of Technology, p.1271-1276.
Häuslein, A., Page, B. (1995) A knowledge-based Simulation Kernel System for the Design of environmental Modeling Tools, in *Environmental Informatics – Methodology and Applications of Environmental Information Processing* (eds. N. Avouris, B. Page), pp. 147-160.
Hilty, L.M., Jaeschke, A., Page, B., Schwabl, A., (eds.) (1994a) Informatik für den Umweltschutz, 8. Symposium, Hamburg 1994. Marburg: Metropolis.
Hilty, L.M., Martinssen, D., Page, B., (1994b) Designing a Simulation Tool for the Environmental Assessment of Logistical Systems and Strategies, in (*Guariso 1994*).
Hilty, L.M., Weiland, U. (1994c) Sustainable Cities – Opportunities and Risks of Information Technology, in *Proceedings of the 13th World Computer Congress* (Brunnstein, K.; Raubold, E. (eds.)) Volume II: Applications and Impacts. Amsterdam: Elsevier. pp. 613-618.
Hilty L.M., Page, B., Radermacher, F.J., Riekert, W.-F. (1995a) Environmental Informatics as a new Discipline of Applied Computer Science, in *Environmental Informatics – Methodology and Applications of Environmental Information Processing* (eds. N. Avouris, B. Page), pp. 1-11.
Hilty, L.M. (1995b) Information systems for industrial environmental Management, in *Environmental Informatics – Methodology and Applications of Environmental Information Processing* (eds. N. Avouris, B. Page), pp. 371-384.
Jaeschke, A., Page, B. (eds.) (1987a) Kolloquium: Informatikanwendungen im Umweltbereich. Kernforschungszentrum 1986, KfK-Bericht 4223, Karlsruhe.
Jaeschke, A., Page, B. (eds.) (1987b) Informatikanwendungen im Umweltbereich. Proc. 2. Symposium, Karlsruhe 1987, Informatik-Fachberichte 256, Berlin: Springer Verlag.
Jaeschke, A., Geiger, W., Page, B. (eds.) (1989) Informatik im Umweltschutz. Proc. 4. Symposium, Karlsruhe, 6.-8. November 1989, Informatik-Fachberichte 228, Berlin: Springer-Verlag.
Jaeschke, A., Kämpke, T., Page, B., Radermacher, F.J.(eds.) (1993) Informatik für den Umweltschutz. Berlin: Springer-Verlag.
Jaeschke, A.(ed.) (1994) Schwerpunktthema Umweltinformatik, it+ti 4/5 94. München: Oldenburg-Verlag, pp. 9-61.
Keller, H.B. (1995) Neural Nets in Environmental Applications, in *Environmental Informatics – Methodology and Applications of Environmental Information Processing* (eds. N. Avouris, B. Page), pp. 127-145.
Page, B. (1988) Environmental Computing - Status and Research Perspectives, in *Computer Techniques in Environmental Studies* (Zanetti, P. (eds.)), Proc. ENVIROSOFT 88, 2nd Intern. Conf., Porto Carras, Greece, Sept. 1988, Southhampton-Boston: Computational Mechanics, pp. 597-607.
Page, B.(1990) An Analysis of Environmental Expert Systems. Environmental Software, Vol. 5/1990, No. 4, pp. 177-198.

Page, B. (1992a) Environmental Protection as a Challenge to Applied Informatics –A Workshop Introduction, in *Education and Society* (Aiken, R. (eds.)), Information Processing 92, Volume II. Elsevier Science Publishers B.V. (North Holland), pp. 595-604.

Page, B. (1994a) Umweltinformatik - Eine neuartige Fachdisziplin der Angewandten Informatik für den Umweltschutz, in *Umweltinformationssysteme in der öffentlichen Verwaltung* (Engel, A. (eds.)), Heidelberg: R.v.Decker's Verlag, pp. 23-47.

Page, B., Hilty, L.M. (eds.) (1994b) Umweltinformatik – Informatikmethoden für Umweltschutz und Umweltforschung. Handbuch der Informatik, Band 13.3. München, Wien: R. Oldenbourg Verlag.

Page, B., Hilty, L.M. (1994c) Umweltinformatik als Teilgebiet der Angewandten Informatik, in *Umweltinformatik – Informatikmethoden für Umweltschutz und Umweltforschung* (eds. B. Page, L.M. Hilty), Handbuch der Informatik, Band 13.3., pp. 13-26.

Page, B. (1995) Database Technologies for Environmental Data Management, in *Environmental Informatics – Methodology and Applications of Environmental Information Processing* (eds. N. Avouris, B. Page), pp. 39-51.

Pillmann, W., Jaeschke, A. (eds.) (1990) Informatik für den Umweltschutz. Proceedings 5. Symposium, Wien, September 1990, Informatik-Fachberichte 256, Berlin: Springer-Verlag.

Radermacher, F.J., Riekert, W.-F., Page, B., Hilty, L.M. (1994) Trends in Enviromental Information Processing, in *Applications and Impacts* (Brunnstein, K., Raubold (eds.)) IFIP Transactions A-52. Amsterdam: North-Holland.pp. 597-604.

Radermacher, F.J. (1991) The Importance of Metaknowledge for Environmental Information Systems, in *Large Spatial Databases* (Günther, O., Schek, H.-J. (eds.)), Proc. Lecture Notes in Computer Science 525, Berlin: Springer-Verlag, pp. 35-44.

Riekert, W.-F., Henning, I., Schmidt, F. (1994) Integration von heterogenen Komponenten des Umweltinformationssystems (UIS) Baden-Württemberg, in *2. Workshop „Integration von Umweltdaten"*, KfK-Bericht 5314, Kernforschungszentrum Karlsruhe.

Rolf, A., Hilty, L.M. (1994) Orientierungen für die Umweltinformatik, in *(Page 1994b, pp. 257-270)*.

Schimak, G., Humer, H. (1994) The Ozone Network for Austria – Technical Concept of a Distributed Environmental Information System, in *Informatik für den Umweltschutz* (eds. L.M. Hilty, A. Jaeschke, B. Page, A. Schwabl), Proc. 8. Symposium, Hamburg 1994, Volume I, pp. 89-96.

Schmidt, M., Giegrich, J., Hilty, L.M. (1994) Experiences with ecobalances and the development of an interactive software tool, in *Informatik für den Umweltschutz* (eds. L.M. Hilty, A. Jaeschke, B. Page, A. Schwabl), Proc. 8. Symposium, Hamburg 1994, Volume I, pp. 101-108.

Simon, K.-H., Jaeschke, A., Manche, A. (1995) Environmental Applications of Expert System Technology, in *Environmental Informatics – Methodology and Applications of Environmental Information Processing* (eds. N. Avouris, B. Page), pp. 93-109.

Sydow A., (1994) Smog Analysis by Parallel Simulation, in *Informatik für den Umweltschutz* (eds. L.M. Hilty, A. Jaeschke, B. Page, A. Schwabl), Proc. 8. Symposium, Hamburg 1994, Volume I, pp. 59-61.

Umweltbundesamt (ed.) (1989) Bibliographie Umwelt-Informatik. Berlin: Erich Schmidt Verlag.

Valk, R. (ed.) (1988) Vernetzte und komplexe Informatik-Systeme. Fachgespräch "Informatikanwendungen im Umweltbereich". Proc. 18. Jahrestagung der Gesellschaft für Informatik (GI), Hamburg Oct. 1988, Informatik-Fachberichte 187, Berlin: Springer Verlag, pp.187-380.
Voigt, K., Bruggermann, R. (1995) Meta Information System for environmental Chemicals, in *Environmental Informatics – Methodology and Applications of Environmental Information Processing* (eds. N. Avouris, B. Page), pp. 315-336.
WCED – World Commission on Environment and Development (1987) Our Common Future, Oxford: Oxford University Press.

7 BIOGRAPHY

Dr. Bernd Page graduated from the Technical University of Berlin and Stanford University and is now a professor for Applied Informatics at the University of Hamburg, Germany where he is doing research and teaching in simulation and environmental information systems. Before his appointment in 1984 he has been working as a scientific associate in the Environmental Information System Group at the German Federal Environmental Office. He received a number of research and development grants and has published extensively in the field of environmental information processing. Dr. Page is co-founder and chairman of the Special Interest Group "Informatics for Environmental Protection" in the German Computer Society (GI) and vice chairman of the IFIP-Working Group 5.11 "Computers and the Environment".

2

Development of an environmental risk characterization methodology for the Philadelphia Energy Risk Study

William C. Miller, P.E.
Penn State University, Great Valley Center
30 East Swedesford Road, Malvern, Pennsylvania 19355 USA
Telephone: 215 685-1225; e-mail: 75051.230@compuserve.com

Abstract

Variations in the emission of arsenic compounds from large power plants in Philadelphia, Pennsylvania, are determined for the period 1950 through 1990. Estimation of the annual average concentration of the airborne arsenic component of particulate matter is presented. A unit risk assessment is used to estimate the historical variation in public health risk. The results of the risk assessment are presented graphically in a contour map format using a "risk power" parameter.

Keywords

Risk characterization, air pollution, power plants, Philadelphia, trace metals, visualization

1 INTRODUCTION

The primary focus of this report concerns the changes in particulate emissions from fossil fuel combustion sources in the electric power generation sector that have occurred through fuel switching in Philadelphia, Pennsylvania (Wainwright). The major fuel switching action to be considered is coal to oil. The effect of this fuel switching action is the variation in the emission of toxic air contaminants.

The primary toxic particulate air contaminants emitted from coal-fired electric utility boilers include the trace metal species arsenic, beryllium, cadmium, chromium, lead, mercury, and nickel (EPA 1991). These seven minerals may also present in fuel oil and are absent from natural gas. The relationship between the type and amount of fuel used and the exposure of individuals to these emissions is explored through public health risk characterization. This relationship can then be analyzed to determine its utility as a basis for linking environmental policy, energy policy, and efforts to understand and resolve the problem of trace metal emissions into the atmosphere (Kowalcyzk et al.).

This linkage would become a useful tool for use by policymakers in formulating emission control strategies. The marketing of various strategy options would be enhanced by the availability of a mechanism to allow the individual citizen to understand the benefit of reduced health risk derived from proposed strategies. This "personalization" process will likely increase the acceptance of otherwise difficult emission reduction measures that, until now, would have been viewed in a detached or impersonal sense.

2 CURRENT METHODOLOGY

An essential aspect of this investigation is the development of an improved method for linking the results derived from estimating toxic air contaminant emissions from large electric utility boilers and characterization the public health risk of exposure to the toxic air contaminants. Fortunately the various individual analytical components required to support this investigation are currently available as described below.

The U.S. Environmental Protection Agency (EPA) has developed a reference database for emission factors for toxic air contaminants as part of the Aerometric Information and Retrieval System - Air Facilities Subsystem (AIRS-AFS). This data is available through the "PM Profile Speciation Report" for each common type of utility boiler. A typical retrieval from this system can provide emission data for individual metal species and other toxic particulate contaminants. General emission estimates are also available in the Compilation of Air Pollutant Emission Factors, published by the EPA (EPA 1991).

The Electric Power Research Institute has developed the Air Emissions Risk Assessment Model (AERAM-2) which assesses the public health risk associated with the emission of air contaminants from utility boilers (EPRI 1985, EPRI 1988). The risk is determined through a four-step process:

1. Characterization of the air contaminants emitted from the power plant.
2. Calculation of ambient pollutant concentrations through air dispersion modeling.
3. Estimation of the exposure of the receptor population to the air contaminants.
4. Public health risk assessment and characterization based on the exposure estimates.

The AERAM-2 emission characterization module contains several options that allow for automatic calculation of emission estimates or the use of actual data such as that provided by the Speciation Report. The Speciation Report data can be further refined where actual fuel chemical analysis is available. The resulting air contaminant emission data are analyzed with meteorological data and receptor location data in the multi-source air dispersion model

ISCLT-2 developed by EPA (EPA 1992). The model results are combined with population data to provide an estimate of the dose to each individual. The dose data and health effects data for each contaminant are then analyzed by dose-response models or analyzed through a unit risk factor analysis.

This general approach is satisfactory for investigating the potential effects of a new source or fuel type change (a *static* situation). It is limited, however, in the study of long-term or historical health effects associated with exposure to airborne toxic trace metal compounds at differing times when using the commonly accepted unit risk analysis method. The key problem lies in the *dynamic* nature of the population (receptor groups) under study. Population is an uncontrolled variable subject to the normal movement of individuals to and from the study area during a multi-period or long-term study. The establishment of a link between exposure and health effect is further confounded by genetics and the variety of exposures that an individual experiences through occupation and personal habits (smoking). This natural phenomenon weakens the usefulness of the population exposure assessment in the AERAM-2 model under these dynamic conditions.

This weakness arises from the nature of the unit risk analytical procedure. Unit risk refers to the specific epidemiological concept defined as the probability of one person developing cancer or some other disease during a seventy year (life span) exposure to one microgram per cubic meter of the contaminant. For example, the unit risk for arsenic is 0.0043 $[\text{ug/m}^3]^{-1}$ (EPA 1995). The application of the unit risk factor is described in Equation 1.

$$Risk = [\text{unit risk factor}] \times [\text{contaminant concentration}] \times [\text{population}] \qquad (1)$$

The numerical values derived from this type of analysis are treated as statistical probabilities. Values less than 10^{-6} are common and are usually accepted as meaning "no effect". Larger values indicate that some level of concern or response is warranted depending on the magnitude and type of health effect (Naugle & Pierson).

This report proposes the modification of Equation 1 by eliminating the population term and converting the resulting exposure risk factor to a "risk power" value, *pR*, as shown in Equations 2 and 3.

$$Exposure\ Risk = [\text{unit risk factor}] \times [\text{contaminant concentration}] \qquad (2)$$

$$pR = -\log(Exposure\ Risk) \qquad (3)$$

This *pR* value permits comparison of potential public health risk at a specific location for multiple time periods without the confounding effect of population dynamics. The logarithmic scale simplifies the display of the large range of data values which may be involved in the plotting of the exposure risk data on a map.

A contour map developed from this type of data can quite easily show the areas of interest by simply designating an action level or "breakpoint" value and formatting the contour line appropriately. For the purposes of the case study described below two breakpoints are defined. Locations where the *pR* value is greater than 6 (equivalent to values less than 10^{-6}) no values are displayed since there would be no effect. Values between 6 and 5 would be displayed, but would be considered a transition region only. Values of 5 or less would be displayed and highlighted by a contour line with a unique font, color, or format. In the regulatory philosophy of many states and the EPA a value of 5 would be considered as the threshold "action level."

3 CASE STUDY DESIGN

The development and demonstration of the proposed methodology is shown through a case study of the history of fossil-fueled electric power generation in the City of Philadelphia during the final year of each decade for the period 1950-1990. Four questions are considered in this case study:

1. What is the estimated risk to public health from the toxic air contaminants (limited to trace metals) emitted from the power plants in the City during this period? The case study and methodology uses arsenic as the example of a trace metal.

2. What are the long-term trends in emissions and air quality due to the operation of electric power generating facilities in Philadelphia.

3. What long-term benefit does fuel-switching provide in a strategy to reduce public exposure to toxic air contaminants emitted from electric power generating facilities?

4. What general environmental and energy policy implications can be derived from the specific results of the case study? These implications may range from local city planning issues to international development issues.

The case study examines the effect of arsenic emissions from the four large electric power generating stations operated by the Philadelphia Electric Company (PECo) within the city limits and the two large PECo stations located in the nearby suburbs. The majority of these generating stations have switched from coal to oil during the study period (Feng).

Exposed population data and corresponding risk estimates are organized on a one kilometer square grid system using census tract based population data. Census data and utility operating represent the years 1950, 1960, 1970, 1980, and 1990 (CPC). Supporting data (meteorology, toxicology, etc.) and analytical software (AERAM-2, mapping, statistical analysis, etc.) are available from public sources.

4 ANALYSIS

Evaluation of the case study and methodology proposed above is presented here in order to demonstrate the feasibility of the proposal. This effort includes the development of the required information resources including power plant physical characteristics, fuel use, emission estimates, geographical and population data sets.

4.1 Geographical Data

The air quality dispersion model used in this investigation requires the establishment of a three-dimensional reference grid system to relate the sources (power plants) and receptors (population) in the estimation of ambient atmospheric concentrations of the subject pollutant. This grid system was created on a scale using one kilometer squares as a representative receptor. The standard Universal Transverse Mercator (UTM) coordinate system is the basis for the grid. The ISCLT-2 dispersion model uses this system as its default horizontal component.

The location of the six point sources for the model and the city boundaries were reduced to the UTM system and define the limits of the model input and output data. This relationship is shown in Figure 1. Source coordinates, city boundaries, and elevations were determined from U.S. Geological Survey topographic maps. A representative elevation in meters was estimated at the center of each grid square. The power plant base elevation was estimated in the same manner. The actual study area (receptor location grid) consists of five hundred seventeen squares covering the entire Philadelphia city limits as shown in Figure 2. The three-dimensional coordinates at the center of each square (UTM east, UTM north, and elevation) locates each receptor for modeling purposes.

The relative location of each major source with respect to the "Center" of the study area is as follows:

Delaware Station	2.6 miles (4.3 km) south
Richmond Station	2.7 miles (4.4 km) southeast
Schuylkill Station	5.7 miles (9.3 km) southwest
Southwark Station	6.3 miles (10.3 km) south-southwest
Eddystone Station	14.7 miles (24.1 km) southeast
Cromby Station	24.3 miles (39.8 km) northwest

4.2 Population Data

The receptor grid system developed above was used to convert census data for the five study years into a population value for each grid square. Census Tract maps for each year were overlain with the UTM grid. The proportion (area) of each census tract within the grid square was estimated. The resulting fractions of each census tract were summed into a population estimate for each grid square.

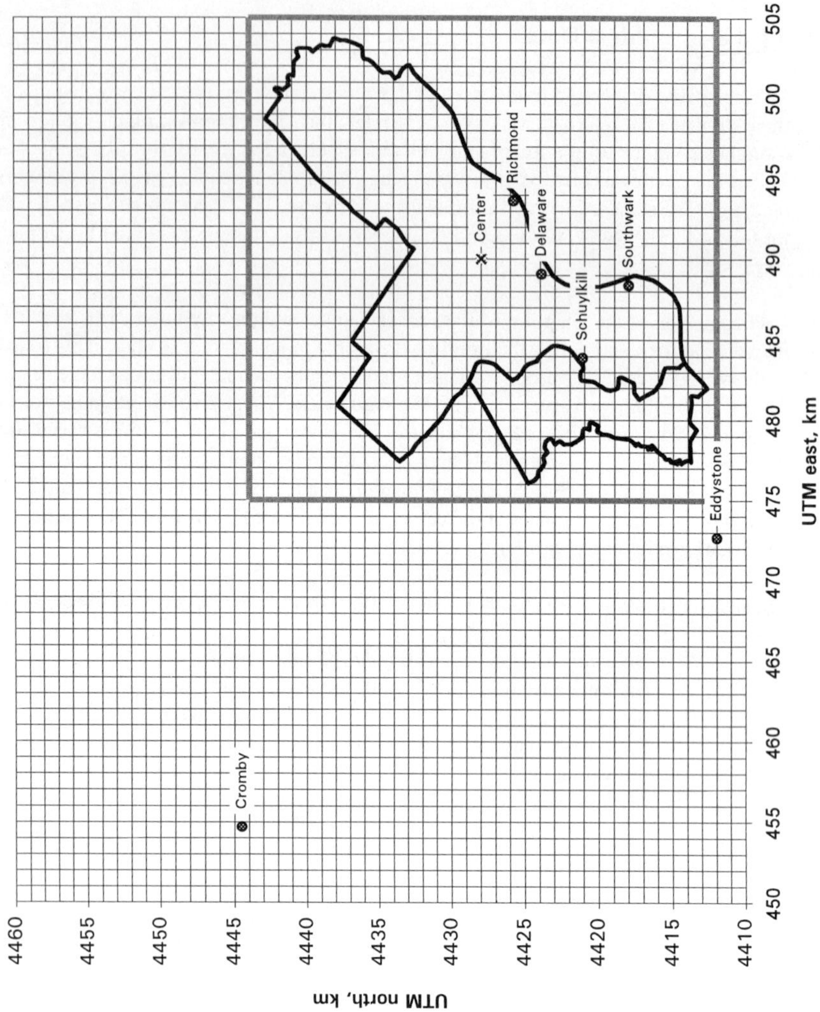

Philadelphia Energy Risk Study

Figure 1. Source Locations and Study Area (Philadelphia City Limits)

Philadelphia Energy Risk Study

Figure 2. Air Dispersion Model Receptor Grid (numbers) and Planning Analysis Sections (letters)

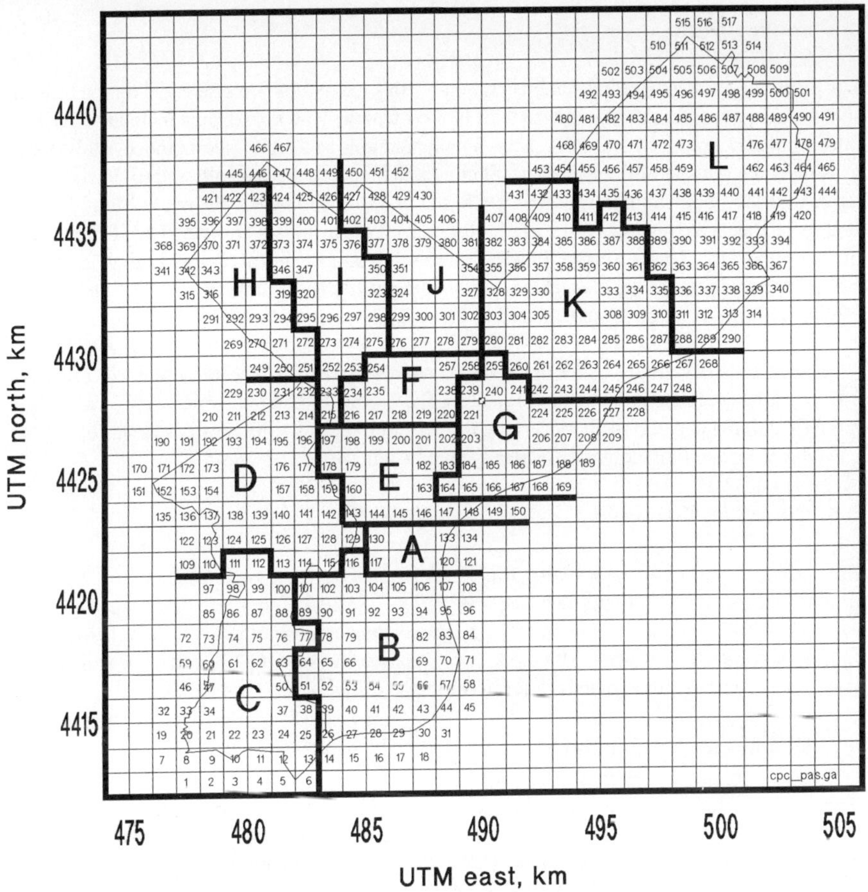

The Philadelphia City Planning Commission (CPC) has grouped census tracts into twelve Planning Analysis Sections for larger scale analytical purposes. A summary of the population variation over the study period is illustrated in Table 1 and Figure 3. This aggregate receptor population grid may be useful in analyzing the results of the risk assessment on a "neighborhood" basis.

4.3 Source Data

The electric power generating stations chosen for this case study were selected on the basis of size, location, tenure, and experience with fuel switching. The six stations included have a rated output in excess of 100 MW each. Smaller stations in the city not included are Willow, Edison, and a portion of the Schuylkill Station which operated seasonally to provide central steam and the small suburban stations at Chester and Barbadoes (Norristown). Monthly reports of fuel type, quantity used, and sulfur content are available from the U.S. Department of Energy and the former Federal Power Commission on FPC Form 67 (or similar) for each station operating under the management of a public utility (FPC 1950, FPC 1970, DOE 1990). This information is summarized in Table 2.

Physical data describing boiler operating parameters and stack size was obtained from the 1971 U.S. EPA emission inventory project for the City of Philadelphia and a 1973 report to the Philadelphia City Council concerning the effect of various sulfur-in-fuel values being considered for the Philadelphia Air Management Code (AMS *a*, AMS *b*).

4.4 Emission Estimates

Fuel use information for each station was disaggregated for each stack (boiler) on a proportional basis using boiler rated capacity. The U.S. EPA *Compilation of Emission Factors (AP-42)* was used to determine emission estimates for Particulate Matter (PM), Sulfur Dioxide (SO_2), Carbon Monoxide (CO), Carbon Dioxide (CO_2), several organic compounds, and the toxic metals targeted for this investigation.

4.5 Dispersion Model Data

The first two modules of the AERAM-2 system described above (source characterization and dispersion modeling) were replaced by the direct use of the U.S. EPA Industrial Complex Source-Long Term (ISCLT-2) model. The model provides receptor concentration results in a file format compatible with the AERAM-2 population exposure assessment and risk characterization modules.

Input to the ISCLT-2 model consists of three types of data. The point source physical data and emission data developed above; summary meteorological data in standard Stability Array summary format (STAR data) representing the long-term weather conditions controlling the

Philadelphia Energy Risk Study

Table 1. Philadelphia City Planning Commission -- Planning Analysis Section Summary

PAS	Section Name	Annual Population				
		1950	***1960***	***1970***	***1980***	***1990***
A	*Center City*	62,373	44,825	44,782	41,897	44,267
B	*South Philadelphia*	297,890	247,863	225,006	185,260	168,105
C	*Southwest Philadelphia*	113,316	103,523	100,248	90,049	85,138
D	*West Philadelphia*	333,749	305,948	279,538	235,922	222,645
E	*Lower North Philadelphia*	368,342	315,351	243,831	176,165	152,186
F	*Upper North Philadelphia*	151,965	139,739	133,990	114,610	105,689
G	*Bridesburg/Kensington/Richmond*	157,220	141,770	126,018	105,004	99,387
H	*Roxborough/Manayunk*	39,035	41,362	50,097	45,896	43,042
I	*Germantown/Chestnut Hill*	101,656	102,892	103,452	89,827	83,709
J	*Olney/Oak Lane*	175,821	189,551	190,126	180,652	172,771
K	*Near Northeast Philadelphia*	237,283	291,793	289,284	253,468	242,088
L	*Far Northeast Philadelphia*	29,955	74,213	161,527	169,462	166,499
	Total	2,068,605	1,998,831	1,947,899	1,688,210	1,585,523

Figure 3. Philadelphia Planning Analysis Sections Population Variation, 1950-1990

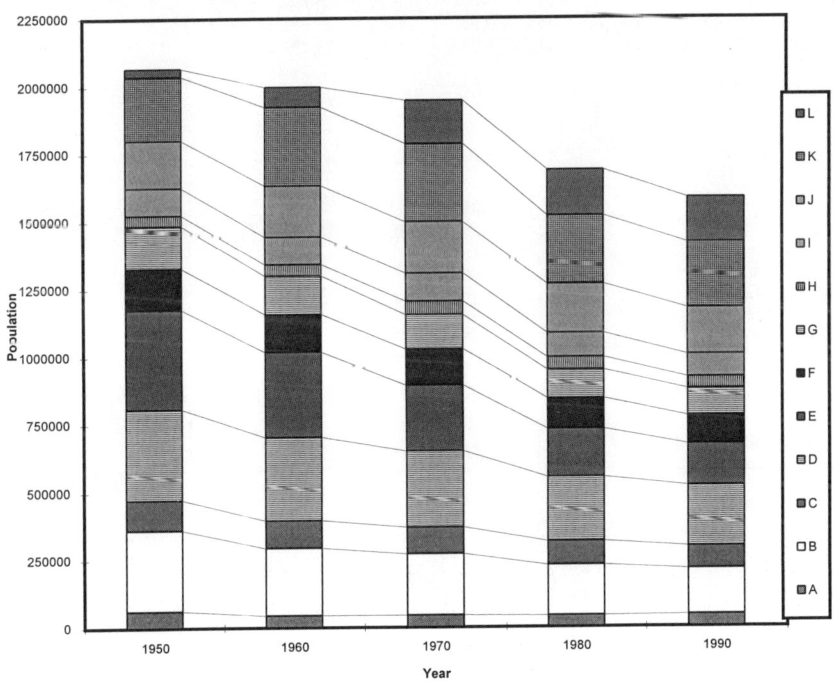

5YR_PAS.XLS 6/8/95

Philadelphia Energy Risk Study

Table 2. Annual Aggregate Fuel Usage by Station

Station	1950	*1950*	1960	*1960*	1970	*1970*	1980	*1980*	1990	*1990*
				Natural Gas Fuel Use (1000 cubic feet/yr)						
Cromby										
Eddystone									1761933	*100.0%*
Delaware										
Richmond										
Southwark										
Schuylkill										
Annual Total	0	*0.0%*	0	*0.0%*	0	*0.0%*	0		1761933	*100.0%*
				#6 Oil Fuel Use (1000 U.S. gallons/yr)						
Cromby					26763	*13.0%*	59206	*15.6%*	17865	*14.1%*
Eddystone							123232	*32.5%*	52584	*41.4%*
Delaware			496	*1.4%*	36047	*17.5%*	44251	*11.7%*	30686	*24.1%*
Richmond			3361	*9.5%*	50874	*24.7%*	17518	*4.6%*		
Southwark	5745	*29.5%*	66	*0.2%*	44842	*21.8%*	32937	*8.7%*		
Schuylkill	13728	*70.5%*	31454	*88.9%*	47594	*23.1%*	102411	*27.0%*	25946	*20.4%*
Annual Total	19473	*100.0%*	35376	*100.0%*	206120	*100.0%*	379554	*100.0%*	127081	*100.0%*
				Coal Fuel Use (tons/yr)						
Cromby			905621	*26.3%*	320385	*18.0%*	287880	*26.5%*	121282	*16.6%*
Eddystone			595016	*17.2%*	1397040	*78.3%*	799932	*73.5%*	610875	*83.4%*
Delaware	252640	*19.2%*	708649	*20.5%*						
Richmond	149217	*11.3%*	667230	*19.3%*	66680	*3.7%*				
Southwark	912386	*69.4%*	573050	*16.6%*						
Schuylkill	845	*0.1%*								
Annual Total	1315088	*100.0%*	3449566	*100.0%*	1784105	*100.0%*	1087812	*100.0%*	732157	*100.0%*
				largest user for year						

FUELUSE2.XLS 6/12/95

dispersion function; and the topographic information describing the three-dimensional coordinates of each source and receptor (population exposure grid system). For purposes of this preliminary investigation a total of five PM and five arsenic model runs were done utilizing all sources and fuel types for each of the five study years.

4.6 Preliminary Results

The results of each PM and each arsenic model run were converted to a citywide summary format shown in Table 3. Using the normal definition of unit risk analysis the risk at each receptor coordinate is the product of the unit risk for arsenic, the annual average arsenic concentration, and the receptor grid population (Equation 1). These summaries clearly demonstrate the effect of the shift from coal to oil in the city and the migration of coal-fired generating capacity to the suburban sites.

Table 3 also indicates the location of the maximum grid concentration and shows the "migration" of the maximum location from the Lower Northeast to West Philadelphia. This long term process follows the rise of coal as the major fuel used from 1950 through the late sixties. In the two year period 1970-1971 a significant number of coal burning boilers were removed or converted to oil. The gradual replacement of generating capacity at older city stations with nuclear power and modern coal and oil burning facilities at Cromby and Eddystone is demonstrated by this same migration pattern.

5 *pR*: RISK POWER

In order to demonstrate the visualization simplicity of the pR variable the concentration and risk data summarized in Table 3 was recalculated according to Equation 3 and displayed geographically in Figures 4a, 4b, 4c, and 4d for the study years 1950 through 1980. The maximum pR grid location is indicated in Table 3 as the Maximum Arsenic pR. The map for 1990 was not significantly different from 1980 and is not shown. The removal of the confounding population variable allows for a clearer view of the effect of the fuel switching and shifting of generation capacity from the older city stations to the newer suburban sites.

The communication of risk information to a wide range of individuals from the scientific community, politicians, and the general public is the most critical component of the risk assessment process. This communication process benefits from a clear visualization method for understanding the phenomenon being studied. Policymakers and individual citizens can see clearly the variation of potential health risk from power plant emissions through the "contour map" style of complex data presentation. The pR parameter presents data having a wide range of numerical values on a scale which is easy to comprehend.

Philadelphia Energy Risk Study

Table 3. Annual Air Dispersion Model Receptor Results

Receptor Coordinates				*Maximum PM Annual Average (ug/m3)*				
X	**Y**	**Grid**	**PAS**	1950	1960	1970	1980	1990
490.5	4425.5	185	G	9.64				
490.5	4425.5	185	G		24.40			
494.5	4426.5	208	G			3.34		
477.5	4414.5	20	C				1.22	
477.5	4414.5	20	C					0.50

Receptor Coordinates				*Maximum Arsenic Annual Average (ug/m3)*				
X	**Y**	**Grid**	**PAS**	1950	1960	1970	1980	1990
490.5	4425.5	185	G	0.0006				
490.5	4425.5	185	G		0.0031			
494.5	4426.5	208	G			0.0006		
477.5	4414.5	20	C				0.0003	
477.5	4414.5	20	C					0.0002

Receptor Coordinates				*Maximum Arsenic Unit Risk*				
X	**Y**	**Grid**	**PAS**	1950	1960	1970	1980	1990
490.5	4425.5	185	G	0.04				
490.5	4425.5	185	G		0.15			
485.5	4419.5	92	B			0.03		
485.5	4421.5	117	A				0.02	
485.5	4421.5	117	A					0.01
City-wide Total Risk				1.57	6.81	2.20	0.72	0.60

Receptor Coordinates				*Maximum Arsenic pR*				
X	**Y**	**Grid**	**PAS**	1950	1960	1970	1980	1990
490.5	4425.5	185	G	5.15				
490.5	4425.5	185	G		4.74			
494.5	4426.5	208	G			5.44		
477.5	4414.5	20	C				5.94	
477.5	4414.5	20	C					6.10

Notes: PM refers to airborne particles emitted from power plant
PAS refers to Planning Analysis Sections (see Figure 2)
Grid refers to a 1 square km model receptor location (see Figure 2)

PMAS_5YR.XLS 6/12/95

Philadelphia Energy Risk Study

Figure 4a. 1950 Airborne Arsenic Unit Risk

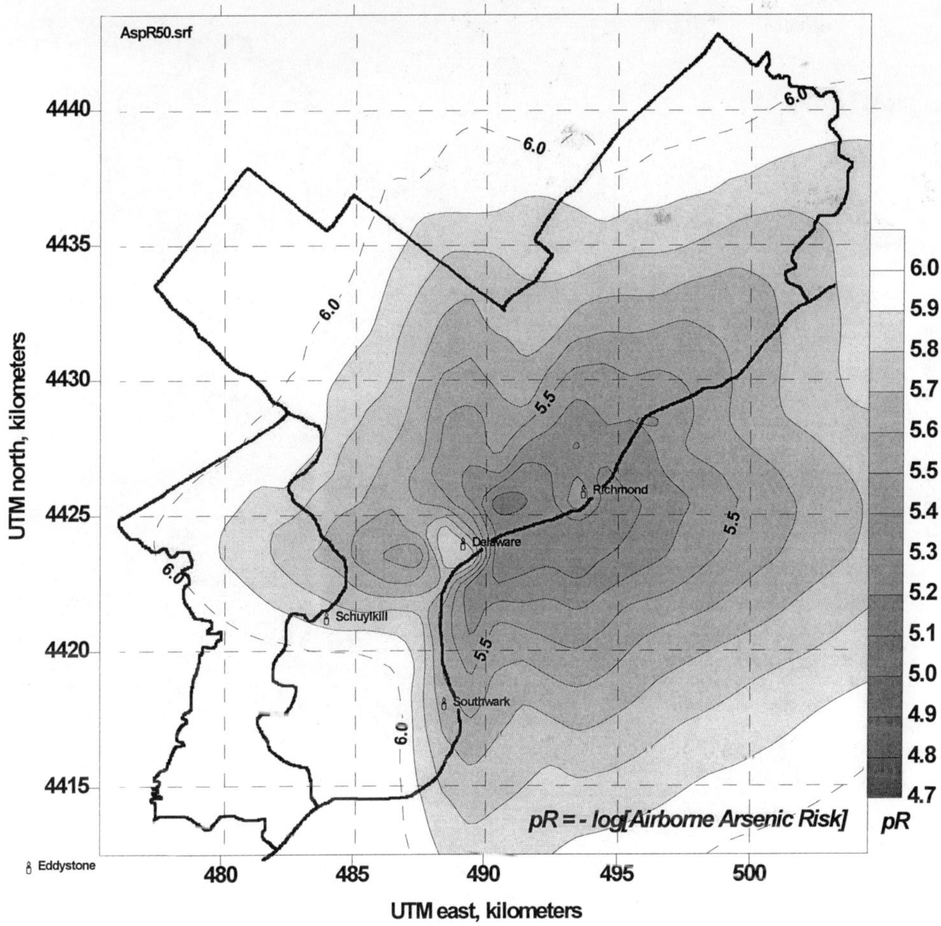

Philadelphia Energy Risk Study

Figure 4b. 1960 Airborne Arsenic Unit Risk

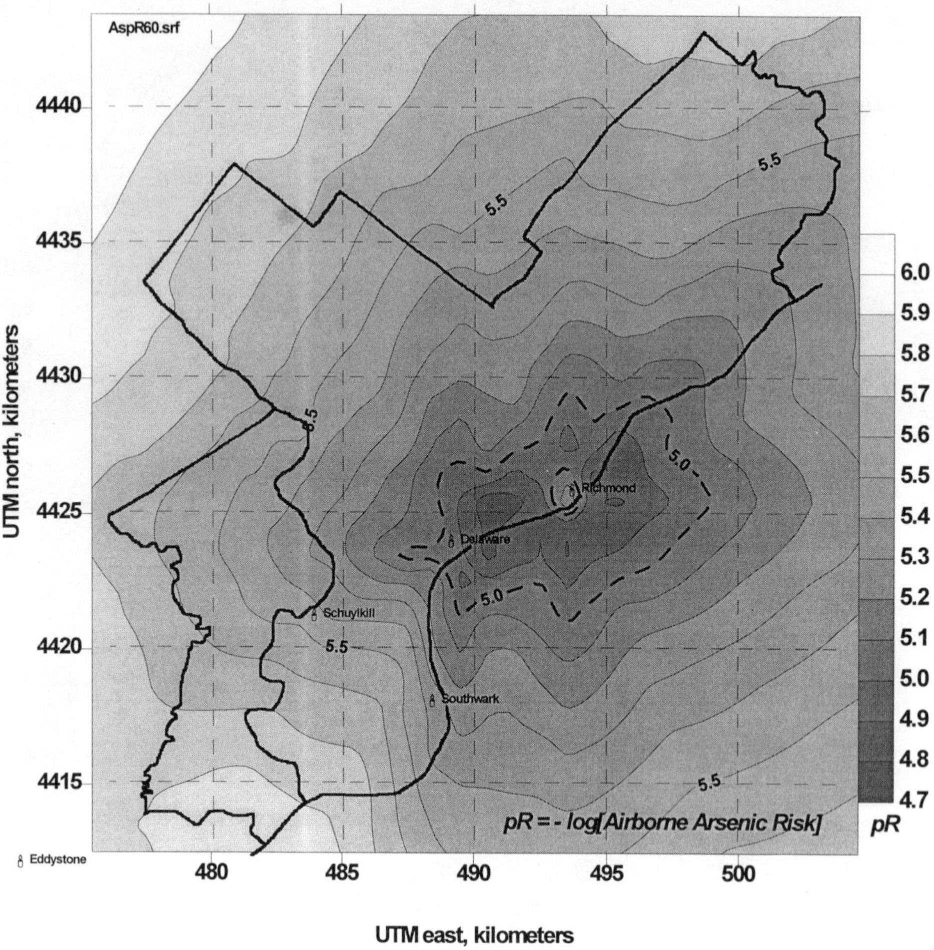

Philadelphia Energy Risk Study

Figure 4c. 1970 Airborne Arsenic Unit Risk

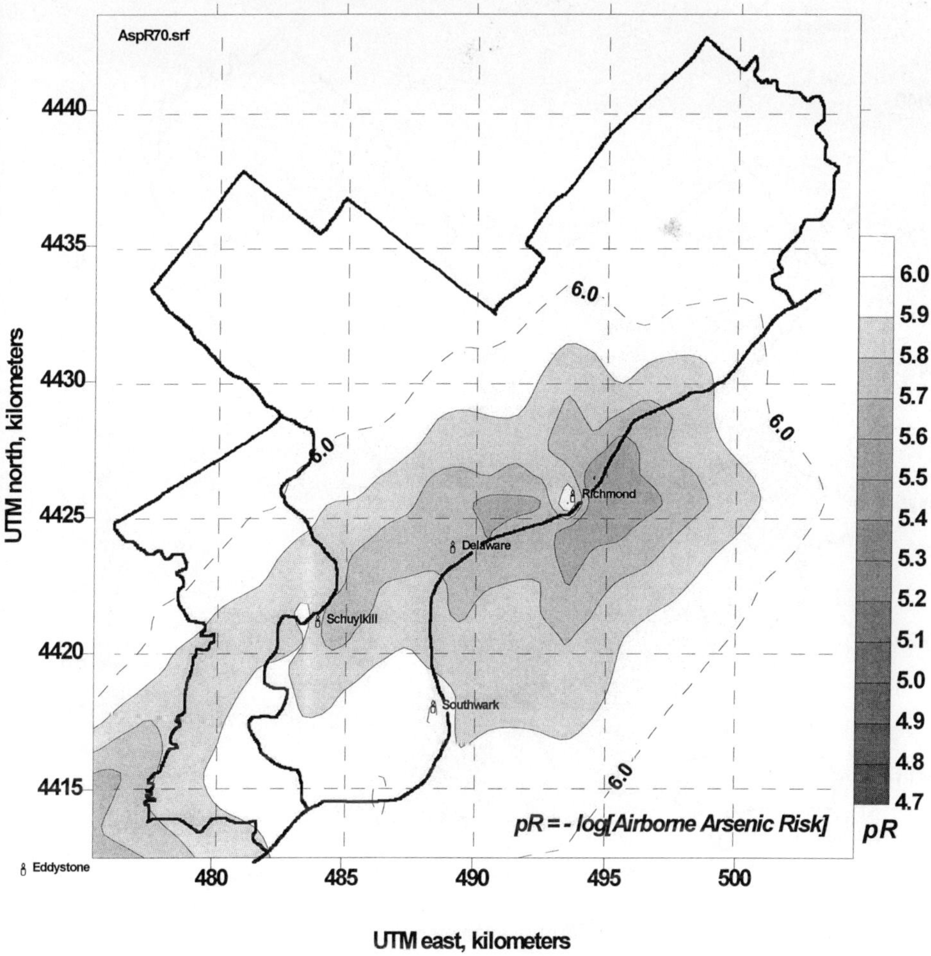

Philadelphia Energy Risk Study

Figure 4d. 1980 Airborne Arsenic Unit Risk

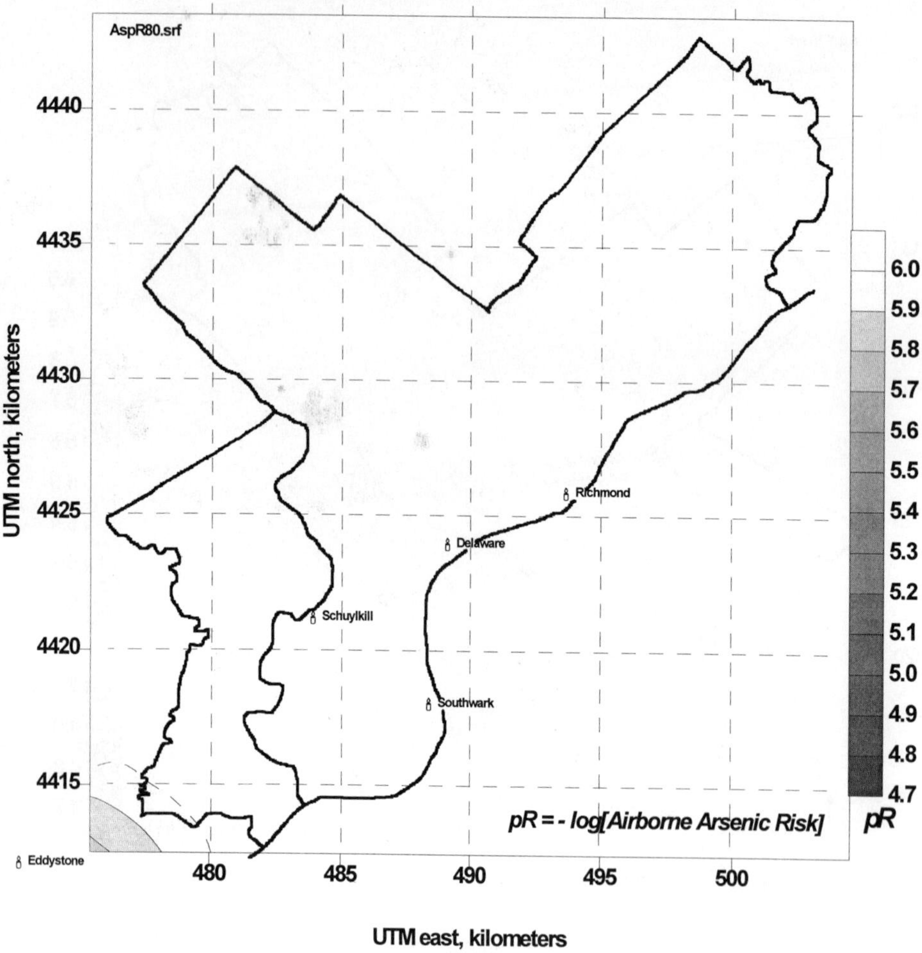

6 REFERENCES

Electric Power Research Institute (EPRI 1985). *Assessing the Health Risk of Airborne Carcinogens*, EA-4021, May 1985.

Electric Power Research Institute (EPRI 1988). *Air Emissions Risk Assessment Model (AERAM) Manager: User's Guide*, EA-5886-CCML, August 1988.

Feng Liu. (1994) "Environmental Risk Modeling," Report to the University of Pennsylvania, Center for Energy and the Environment, June, 1994.

Kowalczyk, G.S. et.al. (1987). "An Air Emission Risk Assessment for Benzo(α)pyrene and Arsenic from the Mt. Tom Power Plant," *J. Air Pollution Control Association*, vol. 37, pp. 361-369, 1987.

Naugle, Dennis F. and Terrence K. Pierson. (1991) "A Framework for Risk Characterization of Environmental Pollutants," *J. Air & Waste Management Association*, vol. 41, no. 10, pp. 1298-1307, October 1991.

Philadelphia City Planning Commission (CPC), Census Reports for 1950, 1960, 1970, 1980, and 1990.

Philadelphia Department of Public Health, Air Management Services (AMS *a*), Annual Emission Inventory Report for 1970, 1980, and 1990.

Philadelphia Department of Public Health, Air Management Services (AMS *b*), Engineering Division, "Electric Power Generating Stations in Philadelphia," 1970.

U.S. Department of Energy (DOE 1990). *Steam-Electric Plant Operation and Design Report - 1990.* Energy Information Agency, Washington, D.C., 1990.

U.S. Environmental Protection Agency (EPA 1991). "Compilation of Air Pollution Emission Factors (AP-42)" , Washington, D.C., 1991.

U. S. Environmental Protection Agency (EPA 1992). *User's Guide for the Industrial Source Complex (ISC2) Dispersion Models; Volume II - Description of Model Algorithms*, EPA-450/4-92-008b, Research Triangle Park, March 1992.

U.S. Environmental Protection Agency (EPA 1995). *IRIS2 Database User's Guide*, Cincinnati, Ohio, March, 1995.

U.S. Federal Power Commission (FPC 1950). *Steam-Electric Plant Construction Cost and Annual Production Expenses - 1938 to 1947; Supplement 3, 1950.* Washington, DC. 1950.

U.S. Federal Power Commission (FPC 1970). *Steam-Electric Plant Air and Water Quality Control Data - 1970.* Washington, DC. 1970.

Wainwright, Nicholas B. (1961) *History of the Philadelphia Electric Company 1881-1961.* Philadelphia, Pennsylvania, 1961.

Software References

Golden Software, Inc., SURFER 5.0; Figures 4a, 4b, 4c, 4d
Hewlett Packard Company, Drawing Gallery; Figure 2
Lotus Development Corp., Ami Pro 3.0; text
Microsoft Corporation, Microsoft Excel 5.0; Figures 1 & 3, Tables 1, 2, 3

7 BIOGRAPHY

William C. Miller is the program adviser for the Environmental Engineering graduate program at Penn State Great Valley. He is the manager for the air monitoring computer center and research programs at the Philadelphia Department of Public Health, Air Management Services Laboratory. Mr. Miller has over twenty-five years of experience in environmental engineering and related data processing systems. He holds advanced degrees in civil and environmental engineering and is a Ph.D. candidate at the University of Pennsylvania, Center for Energy and the Environment.

3

Requirements on Integrated Information Systems from an Ecological Point of View

Lenz, R.J.M.
GSF-Research Centre for Environment and Health, PUC
P.O. Box 1129, 85758 Oberschleissheim, Germany,
Telefon: ++49-89-31872954, Fax.: ++49-89-31873369,
e-mail: lenz@gsf.de

Abstract

The paper intends to give some answers to the following three questions, related to the need of integrated information systems in ecological research: (i) what are the objectives and objects in ecology, (ii) what are the basic sources of electronically supported information in ecology, (iii) what are then the requirements and how could they be achieved? The first question will be answered in subdeviding the „world" in levels of biological organisation of matter and environmental media, for which further classifications and clear definitions are urgently needed. Related to these deficits data and knowledge are still lacking. Besides that, more could be done out of the already existing information if integration and scaling approaches would be enhanced.

The sources for information are basically (relational) database management systems (DBMS), geographical information systems (GIS) and (dynamic) feedback models. In addition, user friendly and cheap data exchanges and accesses to the information systems are needed and to be widely available. Lacking up to now are flexible interfaces to connect the often decentral and specifically designed databases and information systems, and also sufficient meta-informations. In giving some answers to the third question an environmental research information system (UFIS) will be described briefly. It is an operational information system for environmental models, and will be enlarged to data-descriptions in the near future.

Although there are still unsolved technical as well as logical problems, the main need of integrated information systems is the ecological information itself.

Keywords

ecological systems, integration, information system, ecological models

1 INTRODUCTION

The need of integrated information systems in ecological research is quite obvious if we agree in ecology as a new, integrative discipline in the sense of Odum (1977). He urged all the scientific specialities contributing to ecology to merge and integrate holistic and reductionistic approaches in order to meet the challenge of environmental protection. Because of the wide range of objectives of ecological concern one could be doubtful, whether this can ever be achieved. On the other hand, interpretations and applications of ecological knowledge increases drastically, and besides the economics ecology plays an important role in decision making - and thus feeds back ecological research.

To provide the information necessary, a sound basis of knowledge and data has to be elaborated, collected, stored, and retrieved for a broad community of researchers and environmentalists. Data highways are arising, forcing additionally ecology to provide and disseminate their information. Do we have these informations? Are we using the tools already available, and are they sufficient? In spite of the evidence of the need of integrated information systems the basis for that and therefore the reality is rather disappointing.

2 RISING SOME QUESTIONS

2.1 What are the objectives and objects of ecology?

Due to its youth and its wide range of objectives ecology is more diffuse than every other scientific discipline. It is even worse, because integration itself will be only accepted in the scientific community if there is a method behind. And up to now only mathematicians do have distinct rules and methods how to integrate. Odum (1977) didn't mention this kind of methodological lack of integration, but also nobody else.

Ecology deals with interrelations of biota and abiota, organising itself on various levels. These organisational levels of matter (cf. Miller 1975) can be observed, described (structurally) and analysed (functionally). As every part of our environment they are classified objects, integrating pattern and process, structure and function, holism and reductionism. The range of these organisational or integrative levels, the so-called realm of ecology, lasts from organisms to ecosystems and landscapes, or is even more extended. Fig. 1 shows these hierarchy in principle. These levels do have budgets and are closely linked to environmental media like organisms, soils, water, and air. These media are often the objectives of protection and research, but their quality and reactions cannot be understood without understanding the system they belong to. Hence the objectives of ecology are as well the levels of organisation and the corresponding media. Only in distinct and more or less simple cases we can disregard any interaction between them: cause-effects relationship like a shot leading to dead or clear injury are rather the exceptions than the actual impacts on ecological systems and their reactions.

In most cases, clarifying the complex cause-effect relationships is the challenge. Hence it is suggested, to regard three levels of integration in minimum at the same time, if one wants to detect changes or harmful effects of the one level in-between and its interrelation to environmental media (Lenz 1994).

But what does this simple statement really mean? Are these levels finally and operationally defined? And again, how to integrate empirical-descriptive knowledge with the analytical-quantitative one? How to interrelate the knowledge of neighbouring levels? They are mostly not well defined, especially the higher ones like ecosystems, the integration methods are still unclear or very case-specific, and scaling for interrelating the levels was just invented a few years ago!

I will give a few examples, mainly for the ecosystem level, starting with definitions:
An ecosystem is an entity of biotic and abiotic compartments, able to selforganize. Hence the minimum set of parameters describing the ecosystem includes living organisms <u>and</u> interactions with some abiotic environment. Based on this definition an ecosystem is somehow dimensionless and hard to classify. If you do not make further specifications, to my opinion you are lost, as well scientifically (for further analysis) as for application (what to protect?). This is one of the reasons, why Haber (1979) suggested, to embed the ecosystem between biocoenoses and landscapes, or instead, Ellenberg (1973) subdevided ecosystems itselfs hierarchically. But the classification systems are still provisional, and actually a lot of theoretical and task-specific approaches can be collected and discussed (Klijn 1994).

If one agreed in a classification system to be used, the problem of integration still remains. As mentioned before, integration has to deal with combining empirical-descriptive knowledge with analytical-quantitative ones. The original meaning is „to complete something", hence integrative theories, or hypotheses and models have to be delineated from both of the approaches. For example models for forest growth: Statistical and descriptive models like the yield tables, based on the assumption of site constancy, have to be merged with the material balance equation, describing the processes at the site by including the concepts of electroneutrality and material turn over (Ulrich 1987). Both of these approaches alone are still hypotheses and do gain from combining and integrating them, although the process orientation and analysis was a great step forward to come to more reliable predictions.

Interrelating the integration levels means again, to integrate observation with process analysis. For instance, analytical experiments on a landscape level - e.g. changes of landuse patterns which can be related to higher sensitivities like erosional hazards or nitrate leaching - do have their counterparts in observations on the ecosystem level - namely erosion or nitrate leaching! The parameters and models describing the interactions may and will be different at the two levels, but without understanding the landscape the ecosystem analysis will be insufficient because of its therefore limited parameters.

Integration by combining descriptive and process-oriented approaches, and scaling them to neighbouring levels above and below, will lead to more understanding. In addition to the integration defined above, scaling needs to reflect on the level-specific properties and indicate and interrelate them if appropriate.

To conclude these rough answering to the questions arised from ecology, we have to state that integration is a clear necessity but still an unclear method. It depends very much on the definitions of the objectives and the possible harmful effects to be described, analysed and modelled. Therefore the design of information systems must be very flexible and somehow preliminary.

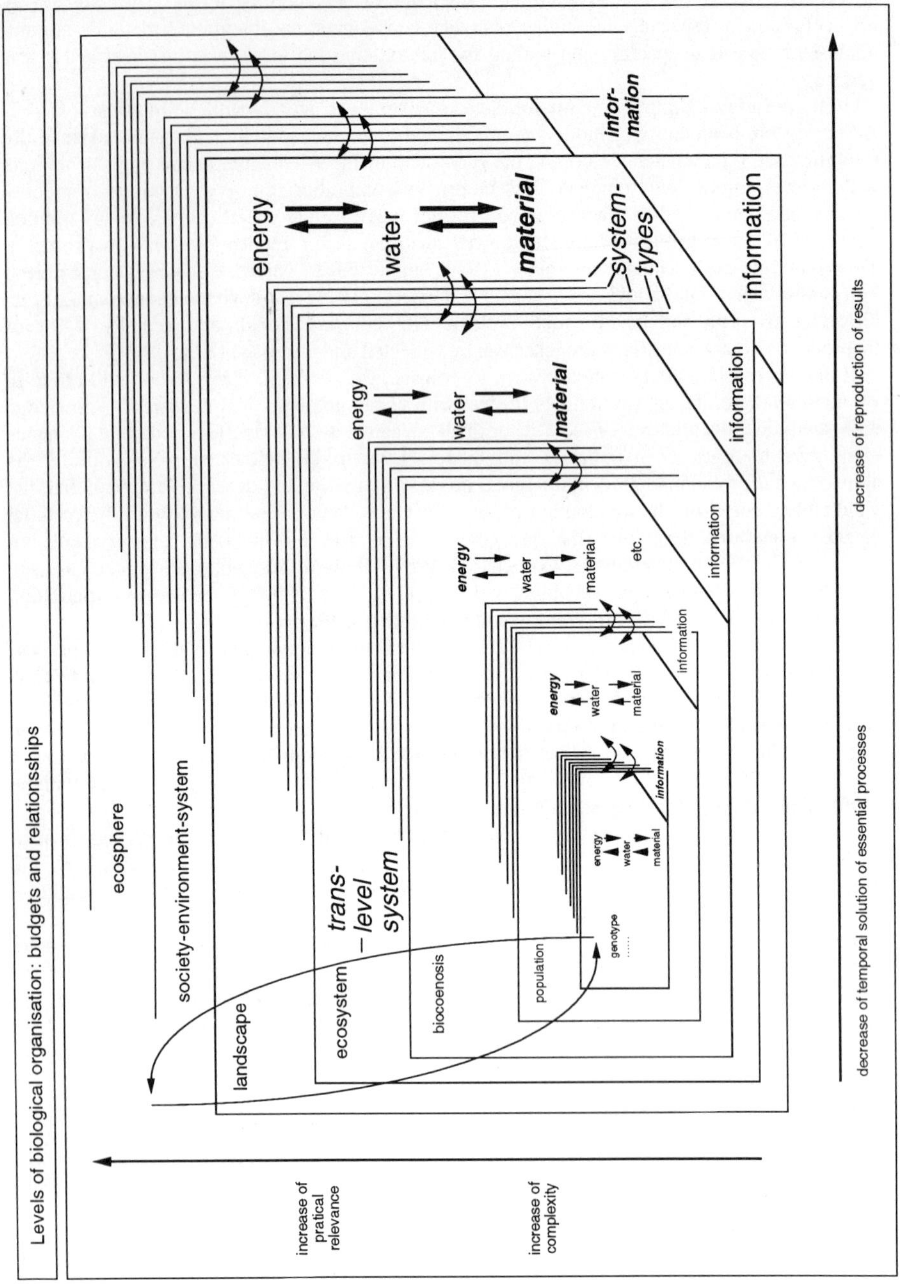

Figure 1 Organisational levels of matter.

2.2 What are the basic sources of electronically supported information in ecology?

The basic sources of information - not only in ecological research - can be devided into three categories:

- point data (and its time series)
- spatial data (and its time series)
- models (to describe the knowledge e.g. at certain integration levels).

Ideally all these sources of information can be combined via flexible interfaces in order to ask for both, models and data, and how they can support each other.
Therefore all these categories do need a very sophisticated detection and description of their context, often called meta information or co-data.
The first category can be covered by database management systems (DBMS), which do provide storage capacities, querying and retrieval. In general their design should not be different for ecologically relevant data, than for other purposes. Nevertheless, the variety of data and the further need of interconnectability do make it necessary to have some conventions on hard- and software used, as well as clear meta-information. In addition, most of the ecological research is performed by non-computer specialists, which means, that user friendly interfaces have to be provided. The same holds true for the spatial data and the GIS and image processing systems. Of course, there are good systems available, but some of them need extensive software knowledge and/or money, and the equipment available for ecological research is far from being adequate concerning these constraints.

The situation becomes even worse, when we look at the facilities of the modelers actually under use. They use what they get, and hence often in very different hard- and software environments. Even the description and documentation is rather incomplete; no routines of plausibility and sensitivity checks are common. As a result the models constructed in most of the cases can be only understood and applied by the modelers itself.

2.3 What are the requirements, derived from ecological needs and information systems actually used?

If we do have to accept the wide range of objectives, the preliminary of classifying the objects, and the also wide range of information systems actually used, we are seriously in trouble.
On the other hand, knowledge and data, information technology and the demand of the scientific community as well as the decision makers for environmentally relevant information increases drastically. In addition there is a need on grading and evaluating these information, e.g. by setting up standards for documentation and meta-information, or even how to gain data and design models.

As a consequence the Ministry for education and research in Germany (BMBF) decided in 1992 to establish an environmental information system, which focuses on ecological research. Up to now the design of the project does not cover all of the aspects ideally to be regarded from an ecological point of view, but is already partly operational. Hence in the following we will give you some insight in the project called UFIS, which is the German acronym for Environmental Research Information System.

3 THE UFIS CONCEPT (SEE LENZ ET AL. 1994)

One part of UFIS is designed to offer information on ecological models as a whole. The goal of UFIS is to provide a tool that supports interdisciplinary work among modellers as well as to give administrative bodies an overview over national and international activities in ecological modelling. The availability of such an information system is believed to be crucial to international co-ordination of modelling activities. Model comparisons and questions of model coupling have been defined as key goals in many IGBP activities (IGBP, 1994). Consequently UFIS will pay particular attention to data requirements of models, areas of application, and scales of the model.

The second part of UFIS which is closely linked to the model documentation focuses on the description of data. The interface to be established follows the items "need of data" and "data yield" in the model documentation. Hence in a final stage of development UFIS will consist of two meta data bases, one on models and one on data.

The background of archiving, describing and the retrieval of such information can be elucidated by reflecting within UFIS the pathways and stages necessary in developing models, and thus boiling down and concentrating the knowledge gained by scientific work. This should make sure that the context of a model - from data requirements to the characterisation of the ecological system considered - is documented within UFIS and therefore retrievable (figure 2). The reason that models and their respective data descriptions were chosen for documentation is that they represent condensed knowledge of complex situations. This does not hold for data, keywords, etc. alone. Here we use the term "model" in the widest sense, basically any kind of combination of data, even simple ones.

The method of modelling can be characterized by five mutually related "sources of information" (figure 2):

1. Conceptual models, i.e. verbal models, hypotheses and theories, strategies etc.. These are derived from observable reality as well as real data about reality (i.e. empirical and quantifying).
2. Mathematical simulation models: they represent the formal counterpart of conceptual models and allow similar as in an experiment to produce data that could be compared to reality. If successful, prognoses and transfer in areas not considered is possible.
3. Computed data,
4. measured data, and
5. observable reality (e.g. certain species and their overall behaviour)

have been mentioned already as required input to and output yielded by models (see above). These five sources of information have their counterparts in the components of the information system UFIS: the retrieval system contains a user guidance by thesauri, systems ecological classifications and environmental media, thereby covering essential parts of the model concept. More detailed information reaching as far as the mathematical description of ecological processes is then accessible from the full model documentation which is available as a paper-based completed questionnaire as well as on-line from the database. This enables identification of data requirements of a model as well as data yield by the model. Note that UFIS contains only data- and model-descriptions (so called meta data), not data and models themselves. The data and models themselves usually relate to an ecological context, i. e. they permit conclusions only under certain ecological conditions. Basically UFIS contains five (or four, if one considers the retrieval system only as a means of access to ecological classification and model documentation) mutually related categories of information covered by the questionnaire. Therefore it is likely that ecological knowledge has been described in a transparent way and qualitatively verifiable.

The systems ecological classification has to cover all ecological systems from the cell (and its environment) up to the ecosphere. The general structure is based on a hierarchical concept of the organisation of matter (Miller 1975, Lenz 1994), on the one hand, and the environmental media (soil, water, air) on the other. Although "ecological" by definition counts for feedback of biotic and abiotic factors, a lot of so called ecological research is classical work in physics, chemistry or biology. Hence the keywording by disciplinary thesauri already in use might be sufficient. Therefore we decided to give further and more flexible orientation in UFIS also by just key-wording what is modelled while not disregarding the ecological classification above. Further descriptions of the context are budgets modelled (if applicable; e.g. water), geographical location, and environmental settings.

Besides the focus on the context of ecological knowledge documented by models and data another feature of UFIS is the flexible degree of details of the information in order to meet the needs of a wider range of users.

3.1 The Model Questionnaire

Model questionnaires have been developed already for specific purposes. There is e. g. one for forest models (Tiktak et al. 1993), and one for crop models (GCTE 1994). These references contain completed questionnaires of 18 and 12 models, respectively. A common feature is that these schemes are tailored for their specific area of application, and they cannot be applied to general ecological models. However, this does not apply to the "handbook of ecological models" (Jørgensen et al. 1994) on which work is currently in progress. This handbook will contain information on several hundred models. All these references give a good sample on the contents of the models but do not provide details like process descriptions. Also the information on the models is not available in a computerised database, but only in printed form. Therefore searches are convenient if one knows the names of the models one is looking for. A severe drawback of all questionnaires that we are aware of, is the freedom that they grant for making entries on it. Most entries are in free text, so the amount of detail given is up to the person who completes it. Furthermore this freedom undermines convenient searches with prescribed keywords. As a consequence we were confronted with the task to design a new questionnaire that is applicable to a wide variety of models keeping the intended use in a computerised database in mind.

The information contained in this scheme on a particular model is threefold: general information, description of processes within the model, and technical information. The first section provides general information on the model. Besides the obvious items like name and address of the modeller, a brief summary of the aim of the model and its basic ideas, valuable information regarding the range of applications is given. This includes references to other models that could possibly be coupled with the one under consideration. The suitability of the model for purposes other than the original one can to a certain extent be judged from reports on calibrations and validations performed (and the data been used) as well as spatial and temporal scales. References to user manuals and to a detailed model documentation are available as well. A flow diagram of the model structure will provide the UFIS user with an idea of the organisation of the model and its sub models.

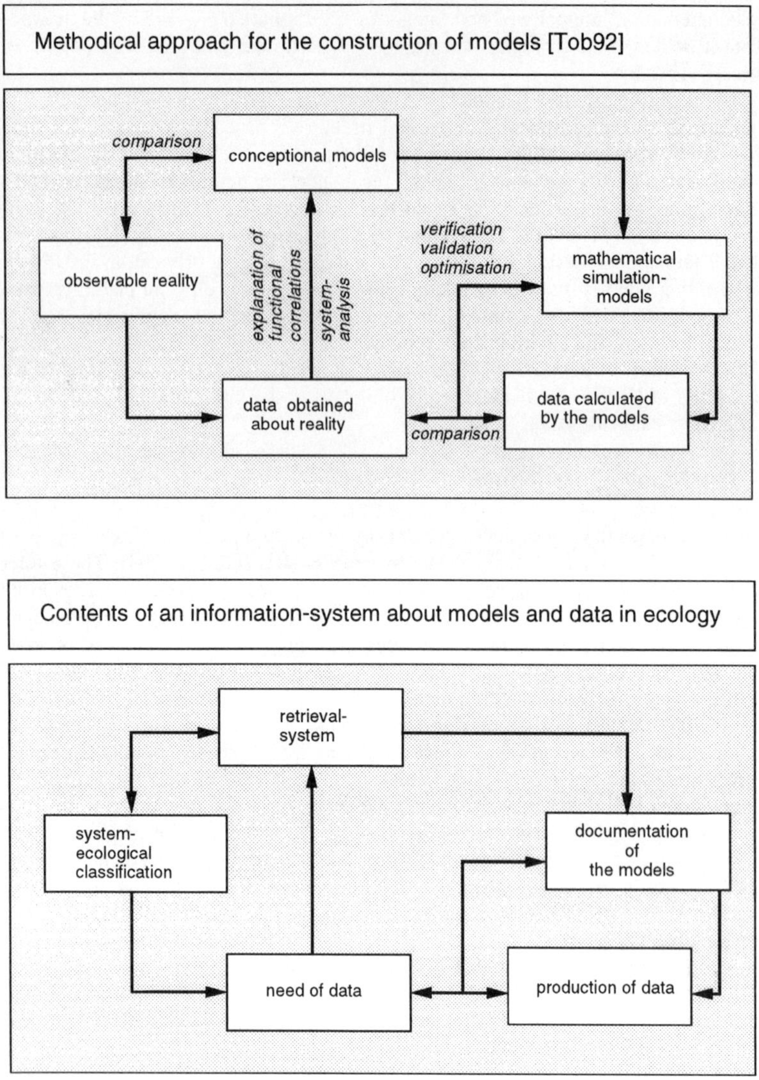

Figure 2 Concepts of modelling versus concepts of information systems.

The second part of the scheme is influenced by the ideas to describe each ecological process and its realisation separately. As the hierarchy among the processes within the model is given by the coupling between these equations via variables information is stored for each process separately, referring to underlying subprocesses via pointers. This information includes for each process: brief description, mathematical formulation, possible constraints, initial/boundary conditions, spatial and temporal resolution, references to the literature and embedded subprocesses. A set of keywords regarding the type of process, type of equation, type of balance/flow enables an efficient retrieval. Quantities used in the mathematical formulation will be classified as *input* and *output* for each process, indicating also which quantities are input and output resp. to the full model. For each of the quantities a brief description, values used, range of ecologically relevant values is given and also the source of the data (where applicable). It is also reported on sensitivity analyses performed by the modeller. In case the mathematical description of a model cannot be split into different processes there is the option to describe the model verbally. However, we designed the structure of the scheme in a very general way, such that this case is considered the exception rather than the rule. For each single process a table of environmental conditions contains a description of its environment. Entries in this table describe the physical environment in which the underlying process formulation is valid. One of the items asked is the "level of organisation" (Miller 1975, Lenz 1993) on which the process takes place and a more specific characterisation within the chosen level (which type of ecosystem, which species etc.). Checking lists for these specifics are supplied; in particular a comprehensive thesaurus for types of ecosystems has been developed. Further items sought are geographic location, type of climate, and type of soil. The particulars are to be given as general as possible without violating the validity of the process formulation. Where meaningful and available, information will be given by means of commonly used classification schemes (e. g. the FAO-scheme for soil types).

The third section of the questionnaire deals with the computer implementation of the model. It is devided in software and hardware related information. Software questions covered are the possible requirement of certain libraries in order to run the model, programming language used and where to get complete program documentation. For the goal of UFIS as outlined above the data question is of particular importance. The UFIS user will find here detailed description of the input files needed and also where to obtain them from. Of course sources of the program code itself, demo input, and sample output files are also given. Hardware requirements are also covered, i. e. for which operating systems model code versions are available, as well as disk space required (Lenz et al. 1994).

3.2 Technical Realisation of UFIS

The goal of the UFIS database development is the coverage of all items from the model questionnaire and their storage in a relational database management system. Therefore the questionnaire represents the foundation of the database. The database is currently being implemented with the database management system Ingres (version 6.3) on a Convex super mini-computer. Usage of the database is possible with very modest hardware requirements via worldwide networks.

Database design:

An analysis of the model questionnaire reveals that it is a strongly structured text document. The items inside the boxes on the questionnaire can each be viewed as a separate page on its own. Within the boxes further subdivisions are possible. These pages together with their mutual relations represent the complete model documentation. Therefore it is natural to view the model documentation as a hypertext document; this is the underlying idea in the design of the database. In order to realise this idea there was in general one table created for each box on the questionnaire. Due to technical limitations some boxes required two tables. Additional tables were also required for the list of keywords and the system ecological classification as well as for the purpose of database administration.

User interface:

There are two user interfaces available for retrieval in the database. The first one is alphanumerically oriented and can be used with any ASCII-terminal. It offers two kind of screens, one for searching and one for browsing. The search screen allows the selection of a set of model documentations by prescribed conditions. The browsing screen provides views on the hypertext pages of the model documentation. Access is via X.25 or internet.

The second user interface is integrated in the world wide web (WWW), a world wide distributed hypertext system. The UFIS database is integrated into the WWW by allowing output of model information in HTML format (hypertext markup language) which is the standard format within the WWW. Access is via URL: http://www.gsf.de/UFIS/ufis.

In addition to the two interfaces for UFIS users, there is another interface for the UFIS managers for administration and input of data. For an illustration of the various interfaces see figure 3.

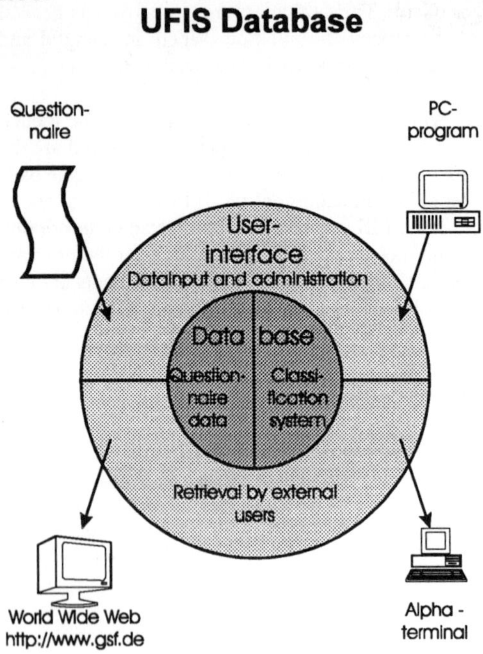

Figure 3 Access to UFIS.

3.3 Perspectives

In this section we want to deal with the questions (i) how to choose models for documentation, (ii) what kind of further interfaces to other information systems should be developed, (iii) how to convince the modellers to document in UFIS, and (iv) how the availability of models and data itself could be improved. The best way to start with the model documentation is not to do this randomly but systematically. Due to the duties and goals UFIS, UFIS started out with BMBF projects and models, selected about 30 from the BMBF-managed database DAVOR, and about 120 by an inquiry at ecosystem research centres throughout Germany. In general future selections will follow the principles of representativeness which means a choice of a wide range of ecological models, both from different levels of organisation as well as more aggregated and more process-oriented ones. At present various information systems are under development, and in the near future a careful check of possible cooperations and interfaces has to be done. An intensive review of information systems in ecological research has been carried out. Also the meetings of UFIS and others supported further links of information systems for adapting the contents or technical features or both. Up to now it is planned to collaborate and possibly interface with the ecosystem research centres, the PIK[1], DKRZ[2], NLB[3], BGR[4], DLR[5], UNEP-HEM[6], GKSS[7], ZADI[8], KfK[9], UFZ[10]and the University of Kassel. The world wide web (WWW) will be the tool for most of these links, but hopefully the documentation scheme will be adopted by the institutions themselves, and many modellers will feed the UFIS database with only little external support. In order to convince others of UFIS and ensure broad support a number of meetings and presentations has been scheduled. Also strong support from the sponsoring agencies is necessary. As can be seen from many other examples scientists and modellers themselves should adopt the concepts and carry out input work with the support by two measures:

- the documentation is part of the project itself ("forcing" the modeller)
- the documentation is widely accepted by the scientific community and leads to closer contacts with other scientists and ecological planners in these fields.

While the first kind of measure is a task of the sponsoring agencies, the second one is ours. Hence there is a need for permanent staff to support the modellers, update and represent the projects.

To improve the availability of models and data themselves is one of the major tasks for the future. We have to come to active libraries of models and data. Starting with the meta-level has the advantage to allow to specify the actual needs and gives a comprehensive overview on

1 Potsdam Institute for Climate Impact Research
2 German Climate Computing Centre, Hamburg
3 Geological Survey of Lower Saxony
4 Federal Institute for Geosciences and Natural Resources
5 German Aerospace Research Establishment
6 UN Environmental Programme - Office for Harmonization of Environmental Measurement
7 Research Centre Geesthacht
8 Centre for Agricultural Documentation and Information
9 Nuclear Research Centre Karlsruhe
10 Environmental Research Institute Leipzig - Halle

models and data that would otherwise possibly remain widely unknown. We should use this advantage for the selection and structuring of true model- and data-bases.

4 CONCLUSIONS

Let me finish my presentation with an example, summarising the needs for integrated information systems. Information is by definition never a single data, its a difference between in minimum two data, may be more, and my be varying ones; and the best information is knowing all the „differences" and contextual data:

Example

5
5kg
5 kg apples
5 kg „ of one tree
" " „ per year
measured with ...
" <the average yield
" but premium quality
" do have a prize of ...>

and: Yield was low because of frost during blooming; growth development can be modelled with the following variables ... but model is only valid for certain environmental conditions and does not include possible pest attacks.

From this example it might be obvious, that neither one database nor a few datas and models are able to cover the field of interest.

5 REFERENCES

Ellenberg H. (1973): Versuch einer Klassifikation der Ökosysteme nach funktionalen Gesichtspunkten. In Ellenberg, Hrsg., *Ökosystemforschung*, Springer, 235-266.

GCTE (1994): Focus 3 Wheat Network: 1993 model and experimental meta data. *GCTE Report*, **2**, Canberra.

Haber W. (1979): Theoretische Anmerkungen zur ökologischen Planung. *Verh. Ges. Ökol. J.*, 19-30.

IGBP (1994): IGBP Global Modelling and Data Activities 1994-1998. *IGBP Report*, **30**, Stockholm, 1994.

Jørgensen S. E., S. Nors Nielsen, and B. Halling-Sorensen (1995): Handbook of Environmental and Ecological Models. CRC Press, to appear.

Klijn, F. (ed.) (1994): Ecosystem Classification for Environmental Management. Kluwer Academic Publ., *Ecology and Environment*, Vol. 2, 293.

Lenz R. (1993): Systemökologische Anforderungen an ein Umweltforschungsinformationssystem (UFIS) als operationale Basis für Scaling - Konzepte in der Umweltplanung. In A. Jaeschke, T. Kampke, B. Page, and F. J. Radermacher, editors, *Informatik für den Umweltschutz*, 7. Symposium Ulm, Springer, 427- 440.

Lenz R. (1994): Ecosystem classification by budget of material; the example of forest ecosystems classified as proton budget types. In F. Klijn, editor, *Ecosystem classification for environmental management*, Kluwer, 117-137.

Lenz R., M. Knorrenschild, C. Herderich, O. Springstubbe, E. Forster, J. Benz, M. Asshoff, W. Windhorst (1994): An information system of ecological models. *GSF-Bericht* **27/94**, 92 S.

Miller G. T. (1975): Living in the Environment - Concepts, Problems, and Alternatives. *Wadsworth Publ.*

Odum E.P. (1977): The emergence of ecology as new integrative discipline. *Science* **195** (4284), 1293-1298.

Tiktak A. and J. M. van Grinsven (eds.) (1993): Model questionnaires. Handout at the workshop on comparison of forest-soil-atmosphere models, Leusden.

Ulrich B. (1987): Stability, elasticity and resilience of terrestrial ecosystems with respect to matter balance. In Schulze E.-D. Zwölfer H., editors, Potentials and limitations of ecosystem analysis, *Ecol. Studies*, Vol. 61, New York, Springer, 11-49.

6 BIOGRAPHY

Education and Degrees

- Ph.D. ("Dr. agr.", summa cum laude) in Agriculture, Munich University of Technology, Germany, August 1991. Thesis: "Characteristics and Loads of Forest Ecosystems in Northeastern Bavaria - a landscape-ecological Evaluation based on materieal budgets" (in German), supervised by Prof. W. Haber and Prof. B. Ulrich
- ("Diplom", summa cum laude) in Agrobiology, Hohenheim University, Germany, June 1982. Thesis: "The impact of wildlife on ground vegetation in forests" (in German), supervised by Prof. J. Pfadenhauer

Experience

- since Sept. 1992: Working group leader "Information Systems for Environmental Research and Planning", altogether about 12 scientific coworkers at the GSF-Research Centre for Environmental and Health
- Jan. 1990 - Aug. 1992: Assistant Professor at the Chair for Landscape Ecology (Prof. W. Haber), Munich University of Technology, Germany
- Nov. 1985 - Oct. 1989: Research scientist at the Chair for Landscape Ecology

4

Software for hard decisions: Scientific influence through interactive visualisation

D. G. Peters
Dept Environment and Land Management, Tasmania
GPO Box 44a, Hobart, TAS 7001, Australia
Phone:+61 (0)02 33 2419, Fax: +61 (0)02 33 3447
e-mail: dp@parks.tas.gov.au

P. K. Robertson
CSIRO Division of Information Technology
GPO Box 664, Canberra, ACT 2601, Australia
Phone:+61 (0)6 216 7087, Fax: +61 (0)6 216 7007,
e-mail: phil.robertson@dit.csiro.au

R. L. Cordy
Dept Environment and Land Management, Tasmania
GPO Box 44a, Hobart, TAS 7001, Australia
Phone:+61 (0)02 33 2419, Fax: +61 (0)02 33 3447
e-mail: rc@parks.tas.gov.au

Abstract

This paper attempts to describe the essential computational functionality required for the understanding and management of the environment and suggests the method by which this functionality might be realised. It is proposed that interactive visualisation will play a key role in enabling the expertise of those involved in the formulation and implementation of government environmental policy, through the provision of new representations and manipulations of environmental information, thus facilitating communication across disciplines and organisational levels in a synthesis of expertise. It is argued that task-oriented user-interfaces will enable access to the advanced computational technologies required to achieve interactive visualisation. Emerging approaches to the systematic design of visualisations and user interaction and the benefits of their application in the environmental arena are outlined. Finally, performance requirements and technologies for realising these benefits, are noted.

Keywords

environmental policy, interactive visualisation, user modelling, ecological sustainability

1 INTRODUCTION

Advanced interaction in the "environmental" application domain

Environmental policy is a major social and political issue - globally, regionally and locally. Policies for resolving of conflicts relating to land and resource use and to standards for environmental protection loom large on most political agendas. Ecological sustainability is generally accepted to be a high priority of responsible government. The operationalising of this high level policy is problematical however.

What benefits can emerging technologies bring to application specialists whose work it is to develop or implement environmental policy? Some technologies have the capacity to be strategic in their use; that is, to allow improved approaches to problem solving, modelling or interpretation. Past examples include geographical information systems, remote sensing, and predictive modelling, the evolution of which have allowed changed approaches to the analysis and representation of land use and biogeography. These technologies have failed to put computers on policy makers' desks in spite of a perceived need for effective decision support. Some of their more serious shortcomings are addressed by "advanced interaction" and its associated technologies.

Advanced interaction is interaction with meaningful data objects and model parameters within the normal working contexts of specialists that is steered by visual feedback. An earlier paper (Robertson, 1994) describes the enabling of specialist's expertise by advanced visual interaction. Much of the basis for the ideas reported here is drawn from that paper.

The aim of this paper is to draw attention to the benefits to nature conservation and environmental management of the application of advanced visual interaction. It is argued that computer systems designed to provide advanced interaction will play a key role in enabling the expertise of those involved in the formulation and implementation of government environmental policy - researchers, planners and managers - through the provision of new representations and manipulations of environmental information, thus facilitating communication across disciplines and organisational levels in a synthesis of expertise.

The purpose of advanced interaction

Unless people can interact with their data, models or hypotheses, it is likely that a significant portion of their expertise is effectively disabled. Learning by observing alone is much less effective than learning when interaction is possible, suggesting that not only can interaction enable expertise but it can also help build expertise.

Visualisations help us to form mental models of data and processes and to communicate our own mental models to others. Mental models are inferred using both the context and the content of visualisations. Interaction can provide the hand-eye coordination that is so crucial to real-world understanding. The process of experimenting by manipulation - adjusting, modifying, or extending - is part of understanding complex concepts and processes. Interaction is also a very effective way of resolving visual ambiguities and inconsistencies in visual representations. Static representations of complex or abstract phenomena are not easily assimilated, but viewers are nevertheless often unable to provide constructive criticism. Manipulation can help to reveal the underlying real world basis of well-designed visualisations.

Advanced interaction within a systematic framework

Systematic design of visualisation, and of interaction, is critical. All too often, visualisations are chosen in an ad-hoc way and fail to illustrate the data or model characteristics of interest. Providing appropriate visualisations and interaction modes will depend on an understanding of how the user expects to carry out the task and on well-defined representations of the task and its products.

A visualisation is generally just part of a broader context, for which there may be many visualisations and interaction possibilities. This "working environment" should be tailored to the user's needs - although usually the reverse is true. Technology, and systematic applications of technology, however, are increasingly offering ways to meet these needs and embed interactive visualisations within the working paradigm of application specialists. Abel et al (1992) describe a prototype of an environmental decision support system which achieved a partial realisation of the goals of advanced interaction.

A systematic approach to the incorporation of interactive visualisation into software systems is essential; as is a systematic approach to developing interaction paradigms consistent with specialists' working environments. Some of today's emerging technologies facilitate these.

2 THE "ENVIRONMENTAL POLICY" APPLICATION

The nature of policy development

Well intentioned, high level policy objectives must be operationalised - reformulated as sets of practical, lower level policies - to have real consequences. Concepts such as biodiversity, ecological sustainability, wilderness, old growth (forests), comprehensive, representative and adequate reservation - the terms in which high level policy is expressed - are complex, abstract and ill-defined. Their assimilation into community mores and the preparedness of industry and government to accommodate them depend on how well the policies they generate reflect credible representations and models of the real world.

The challenge for information technology

Thus environmental policy, on which decisions for land use and environmental protection are based, is a complex, derived product of scientific expertise and perceived public willingness to accept "tough" decisions. For computers to enable policy makers' expertise, they must facilitate the translation of high level conservation and management objectives into real world terms - they must enable the realisation of scientific expertise and its dissemination into practical policy.

Visualisation is already regarded as essential to the development and communication of complex scientific concepts. That interactive visualisation has the potential to multiply these benefits is obvious. The development of systems and interfaces to realise these benefits is an inevitable consequence of our awareness of the importance of these benefits.

A major part of the policy specialist's task is to integrate expertise from many disciplines. Harnessing the power of advanced technologies to integrate information from many sources is a serious research issue (Abel et al, 1994). The difficulty of providing an application environment for this task is much greater than providing it for particular scientific disciplines.

The problem is compounded by a relative intolerance to poor performance while at the same time, data requirements to support this task can be massive because of its dependence on multiple derived data sets and their metadata as well as on primary data. It is clear that an intelligent rather than a brutal approach is required to meet this challenge - it is all too easy to offer indefinite increments in processing power and upscaling of inappropriate software.

In spite of the scale of the problem, the evolution of a productive computing environment for environmental policy development and decision support is necessary. By focussing on their end-use of scientific expertise, and a well-defined user model of policy specialists, within the framework of advanced interaction, useful technology will evolve rapidly.

Examples of tasks

Due to the political importance of the "environment", there are many major policy initiatives with overlapping objectives and integration at this level is an additional overhead, exacerbated by the distribution of responsibility for their implementation among several organisations across all levels of government.

The projects listed do not represent the complete spectrum of environmental policy interests, however, they indicate the range of integration problems and some of the abstract and complex concepts which have to be tackled.

- Wilderness protection
- Rare and threatened species recovery
- Old growth forest conservation
- Ecological monitoring
- State of environment reporting
- Reserve system design
- Sensitivity analysis
- Land use planning
- Analysis of biodiversity

The ways in which the application of advanced interaction might facilitate their handling will be discussed below.

3 INTERACTION NEEDS

Types of interaction

Interaction can range from straightforward pointing or selection, through manipulation and adjusting, to coordinated and highly complex operations. Pointing and selection are well handled by current technologies. Manipulation, however, is much less well handled because of the lack of good input metaphors and supporting devices. Even changing geometrical viewpoint, often the easiest way to resolve ambiguities, generally requires experience with the system being used and often has constraints that are difficult to predict in advance (Felger et al, 1992). This immediately contravenes a core tenet of manipulation in the real world: that we can perform it intuitively or preattentively and allow our attentive resources to be concentrated on observing and interpreting the results of manipulation. With existing computing systems a significant part of the manipulation task requires our attentive mechanisms - just working out how to perform the required manipulation with the tools provided - distracting from observing and interpreting the result. The more successful interfaces for viewpoint manipulation are those that map the interactors to the controls of a real-world learned metaphor, such as driving or flying in a simulator.

Interacting with model parameters - that is, parameters that are significant to the process being investigated - can be a powerful way of building an understanding of model scope and constraints. The direct feedback of the effects of adjusting a model parameter can provide information about non-linearities or other types of behaviour that can be very difficult to determine from static views of the results of a particular set of parameter values. A simple example is the delineation of wilderness quality. Current models of wilderness quality are based on linear combinations of variables such as "biophysical naturalness" and "remoteness". Interactive adjustment of the scaling or weighting of these would almost certainly lead to refinement and provide scope for improving confidence in the output index.

Interacting with model parameters also enables the testing of hypotheses ("what-if" scenarios) with particular benefit to the realism of simulations of landscape processes, such as the spread of fire or pathogens. Simultaneous interaction with both the model parameters and the environmental parameters within which the model is running would be particularly beneficial. This would, for example, allow a modeller to optimise the response of a pathogen model to ad hoc settings of an intervention model's parameters.

A general problem for environmental policy development is the delineation of geographical regions. Reserve system evaluation is often based on biophysical regionalisation because such regionalisations are expected to represent summary patterns in the natural distribution of all the biota. The regionalisation task involves finding a concensus among many individual species distributions. While it can be thought of as a segmentation problem in multidimensional space and hence amenable to computer vision algorithms, these provide only a partial solution (Netravali et al, 1995). This task requires interactive guiding of parameter selection ("computational steering") - to determine suitable thresholding and filtering.

Many environmental policies involve dealing with multi-criteria objectives and thresholds of individual criteria. Interactively manipulating thresholds or the weightings of individual components of multi-criteria objectives would help to detect arbitrariness and would facilitate the evolution of robust indices and flexible criteria.

Achieving interaction - performance and system considerations

These functionalities all require the mapping of model parameters to interaction modes, which means that the model computation must be performed within the interaction latency time. For small problems, computation might be of the order of Mflops (million floating point operations per second), but for models of any reasonable complexity, Gflops (thousand-million floating point operations per second) of computation are likely to be needed. Even this computation size is typical only for screen sized data results. Although processor performance increases steadily, achieving the order of increase in performance needed for interactive modelling requires multiple, or very likely massively parallel, processors.

In fact limitations in screen technologies have allowed, fortuitously, a constraint that has remained pretty well constant for over a decade. Because larger images cannot be displayed, the problem size can often be reduced to that required only for the display resolution. Although not all problems can be so reduced, the constraint did open a window for highly optimised graphics workstations that rely on multiple polygon pipe architectures to achieve the required performance. In general polygon-based algorithms do not scale linearly with data size. Scalable performance needs algorithms that scale linearly with data size and with the number of processors introduced.

These very high performance requirements are thus most likely to be met by parallel processors using pixel-based algorithms where possible (Vézina et al, 1992a,b). Pixel-based algorithms have the additional benefit that they can maintain sampling information, often critical for sensed data from which deductions are being drawn. Because polygon algorithms introduce artifacts (polygon edges) that must then be removed by antialiasing methods to avoid their impact on the display or on subsequent modelling stages, it can be very difficult to maintain and represent accurate original sampling information (Greenberg, 1991).

It should also be recognised that interactive visualisation is probably only going to be helpful if it is integrated within standard working tools, which can pose performance demands outside the scope of those tools. This means improved systems design is necessary (Treinish et al, 1992; Ribarsky et al, 1992; Abel et al, 1994; Sharrott, 1993).

Enabling human expertise

If the performance for interactive visualisation can be achieved, there can be significant benefits by exploiting manipulation and model parameter adjustment. The ability to adjust more than one parameter simultaneously, again something we become very adept at in the real world, can make a substantial difference in exploring the nature of a problem and its potential solutions.

Equally powerful is the potential to exploit interactive forward modelling, and observation of its results, to help solve inverse modelling problems. The specialist is effectively drawn into the process of hypothesising and refining, and in so doing, drawing on experience and learning capacity.

Humans also have an ability to recognise structure or patterns in time varying imagery that may not be evident in static images. Increasing the ability to correlate such patterns with interactively adjusted parameters should more effectively enable analysis skills such as detecting artifacts and surmising their cause, or more clearly distinguishing artifacts from real characteristics.

The environmental policy specialist, who must be able to draw together a variety of complex and incompletely formulated models from many scientific disciplines, is among those who stand to benefit most from interaction.

4 DESIGNING INTERACTIVE VISUALISATIONS

Expertise will only be enabled if the data or processes are presented in appropriate ways that avoid or self-resolve ambiguity, are sufficiently consistent for reproducible results, but are sufficiently flexible to allow exploratory work. This suggests developing systematic approaches to determining the visualisations and interactions that are necessary or desirable, and what visualisations and interactions may be undesirable or introduce ambiguity. Often such approaches are based on a clear understanding, or reference model, of the underlying processes of visualisation and interaction, and the resulting assumptions about the user (Haber, 1990, Robertson et al, 1994).

Choosing visual representations

There are many different alternative visualisations for data, or combinations of data variables, ranging from various ways of using colour to structural representations (Peters (1991) gives some examples of useful ecological visualisations). The question is which is best for a given set of data, interpretation task, and domain of application.

What is needed first is an understanding, in generic perceptual terms, of what characteristics of a data variable are portrayed by any particular representation that is used to convey the information to an observer. This means determining the information conveying capabilities of the various representations used. Robertson (1991) proposes a method for doing this based on visualisations that are depictions of plausible real-world objects or scenes, the so called natural scene paradigm. This paradigm is based on recognising that we are experienced at abstracting and thus understanding intuitively the properties of real scenes, and that this is the same requirement as we have for individual data variables represented within a visualisation - the ability to abstract variables while also looking at their correlation. So for example, an overhead relief-shaded surface representation of data makes evident the local attributes of a particular value (such as gradients), and the global structure (such as undulations), but not necessarily values at a point because there is no absolute reference for surface height. On the other hand, we can introduce an interactive dipstick to give a reading of surface height at any point, or interactively tip up the surface for a perspective view and improve the ability to estimate the height and thus data value at any point.

The natural scene paradigm is particularly suited to the display of environmental information and not surprisingly, draped perspective views are common products of geographic information systems. The natural scene (with its generic point, local and global attributes) need not represent real world terrain. Peters (1991) proposes that "digital attribute models", which are the generic analogues of digital elevation models, should be exploited by biogeographers. They provide a basis for systematically describing attributes of a data variable that can be made evident in any given representation. And in each case, if the data type represented by the scene property, is known, a formalised description of the representation can be made (Robertson, 1991). Examples are environmental scalars such as mean annual solar radiation and precipitation, biodiversity indices such as species richness and habitat quality, and wilderness quality or any of its components.

Understanding the interpretation task

Interpretation tasks are generally couched in domain-specific terms that often have a particular contextual meaning. But that does not preclude the specialist from making the bridge to generic terms such as those used to describe the point, local and global attributes of a data variable that a representation can convey. Usually a specialist has sufficient idea of what is being sought to describe it in generic terms. If the data type is also known, this then provides a formalised description of the data and interpretation task - an important feature for official policy implementation.

Matching the representation to the interpretation

If the representation capability and the interpretation task are both described in generic terms, designing the visualisation and interaction is reduced to a process of matching the representation capabilities to the task requirements, and thus steering the choice of representation. This type of systematic approach to design need not be complex to generate consistency and better user feedback than is achieved without it. Neither does it have to be formalised and implemented in software - it can be a set of guiding steps for realising such representations, with the information about representation capability built up through experience.

The need to develop such systematic approaches to design is being increasingly realised in the area of intelligent information presentation systems: unless there is some systematic base rationale for how intelligence or guidance is applied, it is very difficult to apply it usefully. What is increasingly clear is that with improvements in functionality and performance of visualisation systems, the need for designing visualisations is becoming more evident. The emphasis is very much on moving from pretty pictures to informative pictures, where informative implies that the information can be unambiguously decoded, which in turn implies that the encoding approach must be based on systematic design methodologies.

Visualisations for policy support

The above considerations are particularly important if visualisations are to be used to underpin policy. Another important issue for "official visualisations" is that they must be device independent in the sense that they must have a uniform appearance when displayed on different media. Official hard copy versions are required for reference and must look like the images portrayed on monitors. Also, there is the expectation that policy related material will be published electronically and visualisations should therefore be designed to survive compression and display on popular monitors without loss of information. An example is state of environment reporting in Australia. Methods of satisfying these requirements have been developed (by CSIRO and others) although they are not widely used.

5 INTERFACE AND INTERACTION PARADIGM

The visualisation itself, and interaction with its parameters, are just part of the overall interface requirements. Computing systems should support the normal working environment of application specialists, allowing them to work on problems in a familiar manner, consistent with their normal working paradigm, and make the underlying tools or technologies transparent to their use. Existing computing environments do not currently achieve this.

Working paradigms

Not all environmental policy specialists use computers, but most of those that do, work in an environment that can best be described as a "computational systems and tools" paradigm. That is, they struggle with tools, systems, networks, and incompatibilities. Some work under a "database update" paradigm, but database languages are not sufficient for describing complex modelling or comparison operations and are not easy or natural for policy specialists. A "visualisation-centred" paradigm, based on the recognition that pictures often provide the most effective and influential form of communication, both in static and dynamic form, or even in "hot-spot" enquiry mode sympathetic to an object-oriented software approach, can be an improvement. But even this is several steps away from a "working-application environment" paradigm. Somewhere in between is the map making paradigm of the conventional geographic information system, which is suitable for making maps but difficult to use for anything else

Much closer is a "document or report" paradigm, where the tender of currency is recognised as a document or report for policy advice or positional analysis. An interface that supports such a document as its main component, and where visualisations and other information provide linked access to data and models transparently, can get close to a "working-application environment" paradigm (This paradigm is popular for publishing material on the Internet). But in the end, even this does not properly reflect the actual working environment of policy specialists.

Task analysis and user modelling

Finding the right paradigm needs a more formalised framework. The framework must support the representation and application of knowledge about any particular domain of application, and about users' needs for that domain. It must also support awareness of tasks to be undertaken. Current trials in this area are focused on knowledge engineering - exploiting knowledge representation within a framework sufficiently formalised to capture and maintain relevant information.

A key component is user modelling: that is, modelling and supporting user needs and behaviour patterns, and adaptability to users' evolving needs. Existing software systems, including user interface management systems, focus on systems' functionalities, and what the user needs to do to access them. What is required is the opposite: to concentrate on user needs and wishes, and what the system needs to do to meet those needs and wishes.

Developing an interaction paradigm

This leads to the central concept of developing an appropriate interaction paradigm. Such paradigms can form the basis for designing the interaction the user will perform, and the visualisations on which it will be performed, for a complex series of tasks associated with any given application (Carroll, 1991;True, 1992; Nardi et al, 1993). An interaction paradigm can be viewed as a high level metaphor, embracing groupings of individual metaphors each of which may only be independently sustainable in a limited context within a broad framework that allows the expectations for interaction to be predicted. Such a paradigm should evolve from modelling the user expectations of the system, based on a task analysis.

6 SUMMARY

Technological advances are increasing the computational resources available to application specialists, but often at the expense of increased complexity of use. Current research on designing user-centred interfaces, based on interaction paradigms, offers the potential to change significantly the way in which computing tools are used by focusing on the user and the task, and not on computational tools. The underlying basis of such approaches, drawing on user modelling and knowledge engineering, can provide a systematic framework for developing working environments tailored to suit the needs of any given application field.

Interactive visualisation can enable the expertise of specialists by closing the hand-eye manipulate-view loop and thus allow complex interaction with models and data. This benefit will only be achieved if the visualisations are designed using systematic approaches, again underpinned by knowledge engineering. Advanced visualisation also requires sufficient computational performance to provide interaction with complex model parameters. This performance, several orders of magnitude greater than currently available, will be met by parallel processing technologies supported by transparent software interfaces to application environments.

There seems little doubt that computer systems designed to provide advanced interaction will play a key role in enabling the expertise of those involved in the formulation and implementation of government environmental policy - researchers, planners and managers - through the provision of new representations and manipulations of environmental information, thus facilitating communication across disciplines and organisational levels in a synthesis of expertise. The technologies of knowledge engineering and software engineering are key to realising this benefit.

6 ACKNOWLEDGMENTS

Much of the material for this paper was developed on the basis of a paper presented by P.K. Robertson at Resource Technology '94 in Melbourne, Australia.

The contribution from staff of the CSIRO Division of Information Technology to the ideas behind this paper is acknowledged, as is the influence of discussions with staff of the Tasmanian Parks and Wildlife Service.

7 REFERENCES

Abel, D. J., Yap, S. K., Ackland, R., Cameron, M. A., Smith, D. F., Walker, G., (1992) Environmental Decision Support System Project, *International Journal of Geographic Information Systems* **6**, 193-204.

Abel, D. J., Kilby, P. J., Davis, J. R., (1994) The Systems Integration Problem, *InternationalJournal of Geographic Information Systems* **8**, 1-12.

Carroll, J. (1991) (ed), *Designing Interaction*, Cambridge University Press, New York.

Felger, W. and Schroder, F., (1992) The Visualization Input Pipeline - Enabling Semantic Interaction in Scientific Visualisation, *Eurographics'92, proc. Computer Graphics Forum* **11**(3).

Greenberg D., (1991) More accurate simulations at faster rates, *IEEE Computer Graphics & Applications* **11**(1) 23-29.

Haber, R. B. and McNabb, D., (1990) Visualization idioms: A conceptual model for scientific visualization systems, in *Visualisation in Scientific Computing (ed.* G. Neilson & B. Shriver), IEEE Computer Society Press, 74-92.

Nardi, B. A. and Zarmer C. L., (1993) Beyond Models and Metaphors: Visual Formalisms in User-Interface Design, *Journal of Visual Languages and Computing* **4**, 5-33.

Netravali, A. N and Haskell, B. G. (1995), *Digital Pictures*, Plenum Press, New York.

Peters, D., (1991) Cartographic visualisation and generalisation: Representation of ecological data, *Proceedings: Resource Technology 90, Second International Symposium on Advanced Technology in Natural Resource Management,* 757-765.

Ribarsky, W., Brown, B., Myerson, T., Feldmann, R., Smith, S., Treinish, L., (1992) Object-Oriented, Dataflow Visualization Systems - A Paradigm Shift?, *IEEE Visualization'92*, proc., 384-388.

Robertson, P. (1991) A methodology for choosing data representations, *IEEE Computer Graphics and Applications* **11**(3), 56-67.

Robertson, P. K. (1993), Interactive Visualisation, *Australian Pattern Recognition Society DICTA'93* proc, Sydney, 27-34.

Robertson P. K. and De Ferrari L., (1994) Systematic Approaches to Visualisation: Is a Reference Model Needed?, *Proc. ONR Workshop on Data Visualisation, Darmstadt, 1993, in Scientific Visualization: Advances and Challenges, Academic, 1994*, 287-305.

Robertson, P. K. (1994), Interactive Visualisation: Its role in enabling specialists' expertise, in *Proceedings of Resource Technology '94: new opportunities, best practice;* Melbourne.

Sharrott M., (1993) Visualisation using AVS and Massively Parallel Systems, *AVS Network News* **2**(3), 5-8.

Treinish L. A., Butler D. M., Senay H., Grinstein G. G. and Bryson S. T., (1992) Grand Challenge Problems in Visualization Software, *IEEE Visualization'92,* proc, 366-371.

Treu S., (1992) Interface structures: conceptual, logical, and physical patterns applicable to human-computer interaction, *Intl. J. Man-Machine Studies* **37**, 565-593.

Vézina G. and Robertson P. K., (1992) Data-parallel Visualisation Using Multi-dimensional Transformations, *IEEE Frontiers'92: Frontiers of Massively Parallel Computation,* proc., 230-239.

8 BIOGRAPHIES

David Peters directs the Geographic Information Section of the Tasmanian Parks and Wildlife Service in Australia. He holds an Honours degree in Science from the University of Tasmania. Research interests include the incorporation of artificial intelligence tools with GIS to develop empirical landscape models. Current projects include development of methodologies for terrestrial and marine biophysical regionalisation, biodiversity mapping, forest conservation planning and predictive modelling.

He was the first visiting fellow to ERIN, the Australian Government's Environmental Resources Information Network, where he collaborated with CSIRO on the specification of an environmental decision support system.

Philip Robertson leads the Human Computer Interaction and Visualisation Program at the CSIRO Division of Information Technology in Australia. He holds BEng, and MSc degrees in Electrical Engineering, and a PhD in Computer Science from the Australian National University. His research interests include algorithms and architecture for visualisation of complex data, colour perception and modelling, massively parallel algorithms, and human computer interface. He has been the Principal Investigator for several major projects in the fields involving research and industry groups, and is a member of the editorial board of IEEE Computer Graphics and Applications.

Rosemary Cordy is a researcher in the GIS Section of the Tasmanian Parks and Wildlife Service. She is an experienced public administrator and has provided the strategic direction of the GIS Section since its inception in 1990. She holds a BA from the University of Tasmania and is currently furthering her studies in Behavioural Science at Monash University.

PART TWO

Environmental Information Systems

5

An Information Framework for Environmental Research and Management[1]

D.D. Cowan, P.M. Koch, C.I. Mayfield, D.A. Swayne

*D.D. Cowan, Professor, Computer Science Department,
University of Waterloo, Waterloo, Ontario, Canada,
tel: (519)888-4690, fax: (519)746-5422,
e-mail: dcowan@csg.uwaterloo.ca
C.I. Mayfield, Professor, Biology Department,
University of Waterloo, Waterloo, Ontario, Canada,
tel: (519)885-1211 x4640,
e-mail: mayfield@sciborg.uwaterloo.ca
P.M. Koch, 18 Royal Hunt Crescent, Ottawa, Ontario, Canada,
tel: (613)526-0404, fax: (613)526-0929
D.A. Swayne, Professor, Computing and Information Science,
University of Guelph, Guelph, Ontario, Canada,
tel: (519)824-4120 x3411, fax: (519) 837-0323,
e-mail: dswayne@snowhite.cis.uoguelph.ca.*

Abstract

This paper describes the current status of an ongoing applied research project which is investigating design, implementation and utility of a large scale distributed environmental information system (EIS). This EIS is currently undergoing pilot-scale implementation. A structure for an environmental framework and an accompanying high-level architecture for the EIS are presented. A number of related outstanding Computer Science issues are discussed, and the paper concludes with a brief description of current progress.

---

[1] The work described here has been supported by the National Water Research Institute of Environment Canada, Natural Sciences and Engineering Research Council, IBM Canada, and the Tri-Council Environmental Research Program of the Canadian Government.

Keywords
Environmental information system, environmental framework, environment, client-server system, sustainable management

1 INTRODUCTION

With half the world's population urbanized by the year 2000, sustainable management of urban bioregions is becoming one of the most important social and environmental issues of the end of the millennium. The environmental stresses on urban bioregions are unsustainable in the long term, and basic changes to their management and development are necessary.

Generally, the data describing the ecosystem and the environmental-management systems used to oversee the development of urban bioregions are fragmented, poorly integrated and even at times inconsistent, and a strategic new approach is urgently needed if sustainably managed urban development is ever to become a reality.

At the core of effective urban planning is the environmental and socio-economic information and knowledge to anticipate, and predict the impact of development on a bioregion. Currently, the necessary information, though often extensive, is fragmented, inconsistent, under-utilized and, for many key issues, simply missing. As a result, many of the important cause-effect linkages between the environment and economic development remain unclear. Action to resolve current problems must continue, of course, but what is desperately needed for the future is a substantial forward step in our ability to acquire and use data and information about a given urban bioregion. A common environmental framework is essential for assessing and interpreting regional data, for identifying critical knowledge gaps, for exploring future consequences of policies and actions, and to provide support for education about the environmental relationships in a bioregion.

Development of an environmental framework requires a multi-disciplinary, ecosystem research effort which is supported by an integrated environmental information system (EIS) containing data and knowledge bases in addition to decision-support tools for development planning in urban bioregions. An important consequence of such a system will be the availability of information which can be used as an educational medium for the general public, environmental groups, and political decision-makers.

This long-term, scientifically complex program will require creation of computer-based systems for the acquisition, storage, modeling, analysis, dissemination, presentation and creation of knowledge about environmental issues. Much of the information will be acquired from existing data as well as appropriate field work and laboratory experimentation. A number of computer technologies will be used in this research program, including: data and knowledge bases, geographic information systems (GIS), simulation and analysis systems, expert systems, multimedia and software integration methodologies. Expert systems, multimedia and software integration tools will mold these other computer technologies into a powerful decision-support tool.

Our joint research project which involves a multi-disciplinary group has been operational since 1991 and early versions of the work have been described in (Cowan, 1993a; Cowan, 1993b). The group, representing many areas of environmental research including groundwater modeling and monitoring technologies, geographic information systems, remote sensing, and transportation is determining the environmental framework, identifying interrelationships, and implementing partial prototypes of the supporting information framework. This process has already proven the feasibility of the approach, produced some useful tools, and indicated directions for future research.

2 AN ENVIRONMENTAL FRAMEWORK

An environmental framework is a collection of data, information, and knowledge about the environment in an urban bioregion. This framework is designed to assist environmental planners, engineers, and scientists to create and assess plans which are consistent with sustainable development in a given urban bioregion, to isolate and solve specific environmental problems, and to identify gaps in knowledge relevant to the urban ecosystem. The collection should include geographic and socio-economic GIS data, groundwater data, runoff water data, and environmental impact knowledge to name just a few examples. The environmental framework must be presented in a coherent model so the various relationships among the data, information, and knowledge can be identified, maintained, and exploited.

Much of the environmental data collected today is or can be identified by reference to location or map grid coordinates. Therefore, this is a natural way to relate environmental data sets to each other. Forms of knowledge such as toxicity of various chemical compounds, water quality, and environmental impact assessment rules are not directly related to map coordinates, rather they are used as standards against which the geo-referenced environmental data can be measured.

Since this environmental framework focuses on urban regions and ultimately even neighbourhoods, the environmental framework model must scale effectively as the size of the locale of interest decreases. Thus, the system supporting the framework must be capable of storing and manipulating extremely large volumes of data, knowledge, and relationships between them, and easily expand by allowing the incorporation of new types of data and knowledge as they are identified.

At present environmental information is held by many different public and private organizations that have specific interests in its collection and maintenance. This situation will continue as long as organizations collect information and thus current environmental information will always be distributed both geographically and politically. Clearly the environmental framework is a very large collection of data and knowledge which must be accessed and maintained on a distributed network of computer systems.

3 AN ENVIRONMENTAL INFORMATION SYSTEM (EIS)

The Environmental Information System (EIS) shown in Figure 1 is a distributed information system specifically designed to support the environmental framework. The EIS can be viewed as a client/server system where the client is accessing distributed databases controlled by many different jurisdictions in both the public and private sector. The client supports four major components: modeling, simulation and analysis tools; visualization systems; user interfaces and software integration tools. The client system should also provide support for data access, and the location of new sources of data, software, and expertise.

The entire system is intended to be accessible to environmental planners, scientists, and engineers for their specific research, planning, and problem-solving activities. As well, the information framework must provide educational facilities for environmental groups,

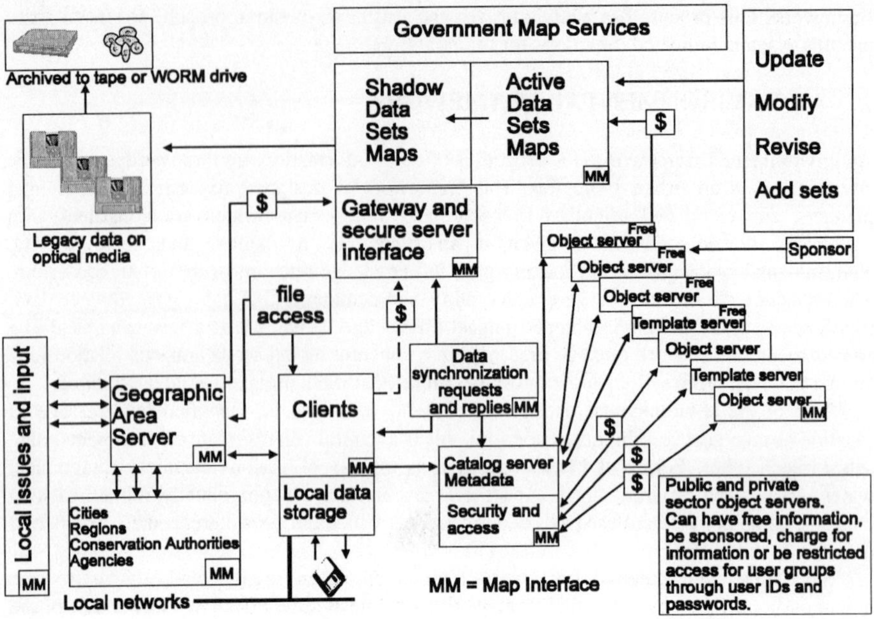

Figure 1 The Structure of an EIS.

politicians, and the public. Since the uses of an EIS are quite broadly based, most applications will be constructed from domain-specific software components using a set of software integration tools. Because the creation of an environmental framework is a research project as well as an environmental planning tool, the components of the system will remain loosely coupled to allow easy replacement of current constituents and insertion of new components.

The servers may supply data, information, knowledge and domain-specific software applications. The "main" database servers will contain base maps showing the primary geographic and physical features of a political region such as a state or province, and will normally be accessible through a government agency. Significant geo-referenced features such as buildings, well data, historical information, or hazardous materials, as well as domain-specific application templates are usually obtainable from local government authorities or organizations such as private consultants, environmental groups or historical societies. Government departments at all levels are cooperating with this project and providing support, since they anticipate that many of the results will help in solving some of their own accessibility and information management problems.

The initial thrust of the project has focused on potential applications of an EIS, since even a partial implementation could have value for a large user community. For this reason, some members of the research team have been focusing on the construction of a prototype client. The client supports local databases and basic tools that have been developed and combined to produce templates for several of the processes associated with environmental planning, monitoring, and visualization including: community-based monitoring, wetlands monitoring, watershed planning, environmental education, and cultural landscape presentation. A number of demonstration systems have been created using these templates. Although the client is not currently connected to the servers, various ad-hoc techniques are used to load and support local databases on the client.

Although the work on developing prototype clients is continuing, we are now focusing on research issues related to accessing the servers over the Internet and prototypes of the Information Highway. The next section describes some of the issues related to this next stage.

4 COMPUTER SCIENCE RESEARCH ISSUES

The information framework which supports the environmental framework must eventually consist of an extensive distributed computer system. Many of the components of the information framework can be implemented now using current computer technology, but a working large-scale EIS will require significant computer science research.

4.1 Databases

The environmental framework will contain data, text, knowledge, graphics, and relationships among these various types. The term database is used in this document as a generic term to describe this entire collection.

Much of the data is structured and describes names of such entities as objects and compounds, and numeric data such as map coordinates, water flow rates, and chemistry. Text will encompass documents which describe specific areas within the map defined by the framework, or various standards which could be used in assessing the environmental health aspects of an urban bioregion. The knowledge will include rules which might describe environmental impact assessment procedures, or water quality tests. Graphic data could be single images, complete video presentations or more complex structures to support visualization of ecosystem behavior. The relationships provide a connection among the various components of the framework, connecting both data to text and knowledge. For example, data on ground water might be connected to water quality assessment rules.

The data in the environmental framework is likely to be collected, maintained and administered by various authorities depending on its relevance to a specific political area. Thus, the aggregate databases will be distributed over a number of computer systems and will be accessible over a computer network. Locating this information and related resources on a distributed database and making such a distributed database system work are challenging research problems. For these reasons, resource discovery (Bowman, 1994), resource browsing (Berners-Lee, 1994) and directory construction (Bumbulis, 1993) are research topics of significant interest.

Once data is located, providing access will still present problems. For example, determining if the correct data set has been chosen for a data analysis or visualization program may require preliminary viewing or computations on a large scientific database of images occupying tens or even hundreds of megabytes per image. Moving this data across the network to the application program can be time-consuming enough to be impractical. One approach under consideration is to provide multi-resolution access to remote data. Preliminary examination or browsing of large amounts of data can be performed using low-resolution data. Tentative analyses can be performed using intermediate-resolutions, and the final analysis can be performed using the data's full resolution. A complementary approach is to reduce data transfer delays, by using the fact that data analysis and visualization programs can often provide substantial information about their future data needs.

4.2 Modeling and Analysis Tools, and Expert Systems

The databases are to be used to support the environmental planning process. In this context it becomes necessary to interpolate and extrapolate the numeric data in the database over space and time to determine likely values for missing data, and to predict the consequences of present actions for the future health of the urban ecosystem. In addition, present data needs to be compared with historical information to observe changes which reflect possible detrimental modifications to the environment. Relationships need to be created between the numeric data in the database and standards so that, eventually, deterioration in environmental data will automatically be brought to the attention of the environmental planner and engineer.

Expert systems and mathematical models exist to assess water quality or monitor land use, and to predict behavior such as the dispersal of toxic wastes in a large body of water, or the effects of acid rain. These are already valuable tools in the environmental planning process. Such tools are still in their infancy and must continue to develop as part of the framework.

For example, the proposed EIS could be used to examine the issues surrounding the exploitation and protection of groundwater resources, since over half the population of North America obtains drinking water from this source. It is important to anticipate future threats to groundwater resources by conducting numerical simulations of the flow of water in the subsurface. The results of such simulations can be used in visualizations to determine optimum placement of landfill sites away from aquifers, and to develop remediation strategies for toxic waste spills. These models require, as input data, a three dimensional description of the geological formation of the subsurface, as well as topographic data, location of streams, rivers, wells and boreholes, well pumping rates, and infiltration data. Currently, development and analysis of the model input is a time-consuming and error-prone task. Much of the data must be obtained from different sources, and re-entered by hand. Addition of new information to these models such as recent borehole test data becomes a tedious process, and hence the models become rapidly out of date. The EIS could eliminate much of this difficulty.

4.3 Computer Graphics

"A picture is worth a thousand words" is an aphorism which has a bearing on environmental planning. Visualization of results in the form of pictures, charts, and graphs, particularly those with some associated animation provides vital support to this process. Graphical systems can also provide a user interface which is more user tolerant and friendly.

There are now many techniques in computer graphics which can produce quite realistic views from computer-generated data and should be available as part of the information framework. New results in graphics and user interface technology should be incorporated into the information framework as they are created. Research in these areas directed toward support of the environment should be encouraged. For example, once simulations are completed, visualization methods could be used to determine the possible extent of subsurface contaminant plumes through examination of fluid flow patterns, or to demonstrate the effectiveness of various groundwater remediation schemes. The Environmental Information System (EIS) could be used to provide widespread access to the resulting high-quality visualizations for both analysis, and to assist in the scientific and political decision-making process. Different bodies such as citizen's action groups and municipal organizations, will want to view visualizations of various simulations in order to verify, understand, and discuss the issues. This will require easy access to large video data sets, and efficient query access to large amounts of simulation data, a problem mentioned in Section 4.1.

4.4 Software Integration Tools

The tools, graphics systems, and databases just described will be used in a variety of ways to support the environmental planning process, and must be configurable to support tasks that are not yet well understood, or even contemplated. This need to use software in an unpredictable and timely fashion means that the various components will have to be loosely coupled, and capable of being "glued" together in different configurations. There is a requirement for a set of software integration tools which will bind together existing components into tools for a new application. Furthermore, the set of software integration tools must allow this binding to occur when the tool is required. For example, tools to monitor and assess a toxic spill cannot wait for the advance planning of the normal software development cycle. Preferably the software integration tools can be used by the application specialist such as the environmental planner with minimal expert help from the programmer. Such tools and the methodologies to use them are starting to become available (Wiederhold, 1992; Cowan, 1992; Cowan, 1993c; NIST, 1994), but significant progress is still required. One approach to this problem is to develop software application templates for restricted environmental processes such as watershed planning, wetlands management, or historical properties' inventory. We are building examples of these templates, which can be configured by the environmental expert to handle specific problems within an application domain. This approach will minimize the need for a programmer except when a new application template is required. Since these templates will use many common presentation, visualization, modeling and analysis tools, we expect to assemble them from various reusable software components created within the context of an application framework for environmental software. Building software applications from reusable parts presents many challenging problems and is a concern of the component-based software initiative in the United States (NIST, 1994). By restricting our work to a single application domain we expect to make significant progress in producing useful tools.

4.5 Networks

Network transmission and switching speeds are increasing dramatically (Aiken, 1992; CANARIE, 1993), and application interfaces are being defined to allow the real-time manipulation and presentation of large volumes of data such as video and other graphic images. Although fundamental network research is not part of this project, certainly network developments will be followed closely, and the EIS will be configured to take advantage of new developments.

5 ENVIRONMENTAL RESEARCH ISSUES

There are two main areas where environmental scientists will have input into the process: data and knowledge evaluation and generation for various environmental models, and research into processes in the environment.

5.1 Data and Knowledge Evaluation and Generation

Data and knowledge already exist for many parts of the environmental framework, but the quantity and quality of this information varies considerably. Approaches need to be developed so that this information can be integrated into the model. The process of integration will assist in identifying gaps in information, and should provide the impetus to create a base standard against which the quality of information can be assessed. One aspect of this base standard will be a set of metadata standards. Metadata will describe the format of the data and its provenance, that is, its origin and quality. A second function of the framework will be to identify and quantify the information needed to continue and enhance the modeling process. Ecosystem level studies are essential in this context since they will lead to superior integration of any information generated.

5.2 Research on Processes in the Environment

Some basic environmental processes are not well understood at present, and yet knowledge of these processes is essential to the maintenance of a healthy urban environment. Examples of these environmental processes include: groundwater effects on surface water, diffuse sources of pollutants and their importance, effects of complex mixtures of pollutants on ecosystems, sequestration and release of materials in sediments and consequences of atmospheric emissions. Conducting experiments, gathering data, integrating data into the framework, and modeling the results in an urban context will provide many opportunities for cross-disciplinary studies in environmental processes. Such research always leads to new questions and new approaches, providing new insights into the ecosystem. In general, accumulation of information into the model will result in modifications to the model through an interactive process.

As the model is developed and new ecosystem research occurs, many new correlations should appear; some will deserve further investigation. Creating an environmental framework should produce a coherent and consistent baseline data set. Such baseline data is extremely important in assessing any future effects observed in the bioregion. By using a standard framework the data will be much easier to access and use, because it will be in a consistent form with standard access procedures. Access standards imply accessibility; data will be available to a wider range of users. Improvements to the environmental review and environmental assessment processes should occur because the model could provide a common database agreed to by major participants in such reviews.

6 STATUS OF THE ENVIRONMENTAL INFORMATION SYSTEM

This paper contains a justification for an Environmental Framework, an outline for an Environmental Information System (EIS) to support the Framework, and presents some of the Computer Science research problems which need to be addressed. We are currently implementing the EIS in an incremental fashion and the remainder of this section contains a description of the progress to date, and some of the problems encountered.

The initial implementation concentrates on the client portion of the system because previous experiences and interviews with many potential users indicated that a properly structured component-based system that could be used to support a wide variety of environmental applications was of fundamental importance. We have constructed a prototype of such a system, and used it to build applications to support environmental assessment and cultural landscape activities. In general we have found that the prototype can be easily configured to support new disparate applications within a few hours, thus giving us confidence in our design and in our choice of basic components as our building blocks. We are currently implementing a more robust version of this prototype, while still using the prototype for continuing experiments.

The prototype version of the EIS now operates over the Internet, in that environmental data such as maps, and map objects and their corresponding attributes can be downloaded from servers to the client system. We have also developed a prototype user interface to allow users to select the data required for their specific applications. Obviously many problems need to be solved as data is acquired from multiple sources. For example, how do we connect map objects which may come from one source with their attributes from possible different sources? Currently such a connection is performed manually. However, we will need to develop some standard approaches to encoding information so that such a process can, as a minimum, be performed semi-automatically. Locating and charging for data are other issues that must be addressed. Various directory structures such as those present in the X.500 standard (Bumbulis, 1993) need to be investigated as well as metadata standards. These and other issues related to the development of an EIS are discussed further in (Cowan, 1995).

7 POTENTIAL CONSEQUENCES OF THE FRAMEWORK

If successful, the process of constructing the environmental framework model should be adaptable to other regions in Canada and to other geographic areas. Third world countries would not have the data available to the same extent, but the model should provide better management decision capability in any given situation. If the framework identifies critical gaps in knowledge in certain areas it will serve an important function. As experience with other areas accumulates, the modeling process will improve. The modeling process can be applied on many scales from very small (university campus), through the municipal or regional level (small towns, cities, metropolitan areas, and watersheds) and could conceivably become a part of a Green Plan for an entire province or country. It is anticipated that it could become a major tool for Planning/Development departments, consultants, public interest groups, governments, research organizations, groups concerned with environmental assessment, and the general public. If access to the information is relatively easy, it would be a significant educational tool for both students and the public.

8 CONCLUSIONS

This paper has examined the structure of an Environmental Information System (EIS) for managing the interactions between human activity and the urban ecosystem. Specific reference is made to current prototype implementations and likely future directions. A complete EIS will require many years and a substantial amount of computer science and environmental research, but the current state of computer technology will allow the creation of an EIS that can have significant impact on our interaction with the environment.

10 REFERENCES

Aiken, Robert. (1992) NSF Implementation Plan for Interagency Interim NREN. gopher://es.net/NREN/NREN documents/impl.txt, May.

Bumbulis, P.J., Cowan, D.D., Durance, C.M. and Stepien, T.M. (1993) An Introduction to the OSI Directory Services. *Computer Networks and ISDN Systems*, 239-249.

Bowman, C.M., Danzig, P.B., Manber, U. and Schwartz, M.F. (1994) Scalable Internet Resource Discovery: Research Problems and Approaches. *Communications of the ACM*, 37(8):98-107, August.

Berners-Lee, T., Cailliau, R., Luotonen, A., Nielsen, H.F. and Secret, A. (1994) The World-Wide Web. *Communications of the ACM*, 37(8):76-82, August.

CANARIE. (1993) CANARIE: The Canadian Network for the Advancement of Research, Industry and Education. CANARIE Inc., 401 Laurier Avenue West, Suite 1120, Ottawa, Ontario, K1P 6H5.

Cowan, D.D., Ierusalimschy, R. and Stepien, T.M. (1992) Programming Environments for End-Users. In *Proceedings of IFIP 92*, Volume III, 54-60.

Cowan, D.D, Grove, T.R., Mayfield, C.I., Newkirk, R.T. and Swayne, D.A.B. (1993a) An Integrative Framework for Environmental Management and Research. In *Proceedings of the Second International Conference/Workshop on Integrating Geographic Information Systems and Environmental Modeling*, 6 pages.

Cowan, D.D., Grove, T.R., Mayfield, C.I., Newkirk, R.T. and Swayne, D.A.B. (1993b) Managing Environmental Data An Extension to the GIS Architecture. In *Proceedings of the Workshop on Advances in Geographic Information Systems*, 10 pages.

Cowan, D.D., Ierusalimschy, R., Lucena, C.J.P. and Stepien, T.M. (1993c) Application Integration: Constructing Composite Applications from Interactive Components. *Software Practice and Experience*, 23(3):255-276, March.

Cowan, D.D, Koch, P.M., Mayfield, C.I. and Swayne, D.A.B. (1995) Implementing an Information Framework for Environmental Research and Management. In *International Proceedings of the Conference on Computing and Information*, 10 pages, IEEE Press.

NIST. (1994) ATP Component-Based Software Program. gopher://zserve.nist.gov:79/1/.menus/.atp94/.9406.

Wiederhold, G., Wegner, P. and Ceri, S. (1992) Towards Megaprogramming. *CACM*, 35(11), November.

11 BIOGRAPHIES

Donald D. Cowan received a B.A.Sc. degree in Engineering Physics from the University of Toronto in 1960, and the M.Sc. and Ph.D. degrees from the University of Waterloo in Applied Mathematics in 1961 and 1965, respectively. He has been on the Faculty of the University of Waterloo since 1962 and is currently a Professor in the Computer Science Department and Director of the Computer Systems Group. His research interests include software engineering, and the design and implementation of computer systems to support environmental management and planning. Dr. Cowan is a member of the IEEE Computer Society and the ACM.

Paul M. Koch received the B.A.Sc. and M.A.Sc. degrees in Chemical Engineering from the University of Waterloo in 1963 and 1964, respectively. He has been a member and Vice-Chairman of the University's Board of Governors. His interest in environmental issues developed from attendance at the UN Conference on the Human Environment in Stockholm, in 1972. He spent the majority of his business career in marketing assignments with IBM Canada Ltd. Since 1992, he has been self-employed and specializes in developing and facilitating business strategies that use new approaches to education and the use of information technology in handling environmental issues.

Colin I. Mayfield received a B.Sc. and a Ph.D. in Microbiology from the University of Liverpool in England in 1966 and 1969, respectively. He has been on the Faculty of the University of Waterloo since 1970 and is currently a Professor in the Biology Department. His research interests include environmental toxicology, the bioremediation of contaminated groundwater, the design and implementation of computer systems to support environmental management and planning.

David A. Swayne received a B.Sc. and Ph.D. from the University of Waterloo and an M.A. from York University. He is currently a Professor in the Computing and Information Science Department and the School of Engineering at the University of Guelph, and an adjunct professor in the Computer Science Department at the University of Waterloo. His research has focused, since 1983, on applications of computing science and information technology to the environment. Since 1991, he has collaborated with the University of Waterloo's Environmental Information Systems Project. He is a member of the Society of Industrial and Applied Mathematics.

6

Semantic enhancement of forest inventories by reengineering

H. Hasenauer[1)] *G. Pernul*[2)]
[1)] *Institute of Forest Growth and Yield, Department of Forestry, University of Environmental Sciences, Peter Jordan Str. 70, A-1190 Vienna, Austria.*
E-mail: hasenau@edv1.boku.ac.at
[2)] *Institute of Applied Computer Science and Information Systems, University of Vienna, Liebigg. 4/3-4, A-1010 Vienna, Austria.*
E-mail: pernul@ifs.univie.ac.at

Abstract

In order to enhance the semantic representation of data collected in forest inventories the paper proposes a reengineering technique consisting of Entity-Relationship modelling and a transformation process into relational database structures. The work is motivated by the fact that a huge amount of information about our most complex ecosystem on earth, the forest, is available but unfortunately currently only interpretable by the forest experts. This is due to the current structure of the data and in our opinion this is a huge disadvantage because data collected may also be of tremendous value in related disciplines, for example, for solving ecological questions. Enhancing the semantics and providing a more convenient user interface for accessing the data may make the information available in a much easier way to the ecologist, the botanist, the wildlife expert, and to all other parties that might be interested in using the inventory.

Keywords

Data modelling, reverse database engineering, forest inventories, semantic enrichment

1 INTRODUCTION

Forest inventories are carried out by forest companies for collecting information about their forest lands in order to develop efficient forest management plans. Another type of forest inventories are implemented by the national forest experimental stations of European countries (for example, in Austria, Finland, or Germany) with the main purpose to estimate and control the forest covered land areas of their countries and to observe the state of forests under changing environmental conditions.

Today most forest inventories are computerized but are mainly designed and used by forest experts and the data collected are generally coded and stored as simple ASCII files. For extracting information several application programs must be developed and applied to the data.We will refer to such a file structure as *flat files* in the following.

While for certain forestry applications (and also for the forest engineers) the structure of the data has proven over the recent years to be sufficient newer applications, for example, software support for environmental protection, ecosystem software, or ecological management software can only make limited use of the data because of its current structure. This is a huge disadvantage because the data collected may be of tremendous value for solving ecological questions. This is because forests are considered as one of the most complex and highly developed ecosystems and thus react to changing environmental conditions rapidly. From the viewpoint of an ecologist, who unfortunately rarely is also an expert in forestry, reality may be seen differently from what was considered by the forest engineer at the time the structure of the forest inventory was determined. It might be a realistic but the worst case that it is the opinion of the ecologist that data has a poorly designed structure, applications are poorly coded, use poor logic, are poorly documented and for him as a consequence the whole forest inventory does appear as not representing the semantics of the application domain properly. This situation is a very good example of the fact that it is not possible to get a system permanently complete, because it is not possible to predict now what we will want the system to do ten, twenty, or even thirty years from now.

As a solution to the problem discussed above we suggest to reengineer existing forest inventories in order to make information about our forest available and better interpretable for non forest experts. Reengineering is a research discipline within Software Engineering with its main purpose to carefully analyze a given system, to identify the system's components and their interrelationships, to create another - more advanced - form of the system at a higher level of abstraction (i.e. to reverse engineer the system), and finally to use the abstractions to reconstitute the system in a new semantically enriched form. As a result the new version of the system should be able to better meet the new requirements in comparison with the old form of the system.

The objective of this paper is twofold: 1) We develope a robust methodology to reengineer information stored as flat file structures in existing systems. The methodology uses state-of-the-art data modelling and database design concepts in order to enhance the semantics of the new form of the system and to reduce redundancy, inconsistencies, and system maintenance costs. 2) We report our experiences in using this methodology in a case study on the reengineering of one of the most complex forest inventories in Europe - the Austrian National Forest Inventory, OEFI (1994). A prototype application has been developed in MS Access to test the adequacy of the proposed methodology.

The outline of this paper is as follows: In the remainder of Section 1 we discuss related work. In Section 2 we describe the current form of the forest inventory used as the starting point of our efforts. In Section 3 we explain the methodology developed to reengineer flat file structures. For its understanding some basic knowledge of data modelling and the relational data model is necessary. Section 4 is devoted to the case study. In Section 5 we conclude our work and discuss open research questions.

Related Work

Reverse software engineering is a recognized research topic in which a huge amount of work has been published recently. Instead of referring to particular articles we find it more useful to point only to three special issues of prestigious periodicals devoted to reverse software engineering, the CACM (1994), Software (1990), Software (1994). Articles appearing in these special issues can be consulted to gather further references. Our work is also related to conceptual and logical database design. For conceptual database design the Entity-Relationship (ER) Model of Chen (1976) and its variants serve as a *defacto* standard and were also chosen as the underlying conceptual data model for reverse engineering in our approach. Database design in general is treated in detail, for example, by Navathe and Pernul (1992), by

Batini *et. al* (1992), or by Theorey *et. al* (1986). Although considered as an important topic methodologies for reengineering of databases are not reported in the literature frequently. Navathe and Awong (1987) propose a systematic approach for the reverse engineering of relational databases into extended ER models, Markowitz *et. al* (1989, 1990) focus on the more theoretical aspects of reverse database engineering and Premerlani and Blaha (1994) use an object-oriented design and analysis for the reverse engineering of relational databases. We are not aware of any other project devoted to the reverse engineering of forest inventories.

2 THE APPLICATION ENVIRONMENT

The application environment used in this study is given by the design and data collecting concept of the Austrian National Forest Inventory. The tasks and the data collecting system as it is shortly described below is well documented in a manual called "Instructions for the field work of the Austrian Forest Inventory", OEFI (1994).

The general objective of the OEFI is the recording of the Austrian forest covered land, the timber growing stock, the increment and the felling quantity. In the sixties the sample plot system was introduced and currently the OEFI consists of about 10.000 permanent sample plots and every year a fifth (five year measurement cycle) of the plots spread all over Austria are remeasured. During the last decade the tasks have been extended rapidly and today a huge amount of data are recorded on a plot and tree basis for describing site and stand characteristics, stand structure and treatments, regeneration, damages by game, and which is of increasing interest - variables for characterizing the health conditions of trees and stands.

Because a permanent plot system was installed always the same plot location and even the same trees on a plot are observed. The advantage of such a system is that the current status can be compared with previously ones and so the change itself can be considered directly.

The data collected can distinguished into plot and single tree related variables. The *plot related information* is collected on a fixed plot size of 300 m^2. Related variable groups describe some general plot information, like the percentage of forest covered land, land owner, political district etc., common stand data, like stand structure, age group, growth class, growth district, etc. Furthermore, a very detailed information is given about the site conditions of the plot by considering the soil and humus conditions, the steepness, the elevation, the exposition, etc.

The recording of *single tree related information* is dependent on two different plot size: For trees with a diameter at breast height (DBH) of more than 10,4 cm the angle count system with a basal area factor of four is used for selecting the sample plot belonging trees. The angle count system gives thicker trees a higher probability to be included in the sample than thinner ones because they contribute a higher percentage of the basal area or volume per unit area and should therefore be measured more accurately. From the statistical point of view it is important to note that every tree has its own sample plot size depending of its DBH. For trees with a DBH between 5,0 - 10,4 cm again a fixed plot size of 21 m^2 is used. Trees with a DBH of less than 5 cm are not included.

For every tree the location on the plot is defined by measuring the azimut and the distance between the tree and the permanently marked plot center. Tree species, DBH, total tree height, height to the base of the live crown, etc. are determined. Furthermore the tree class for characterizing the tree's competition situation and some variables describing the size of tree damages like stem and root damages, number of bark peelings, the crown condition and the needle fall are given.

Finally some general economical information concerning the expected wood quality, the cutting and logging conditions, like distance to the next road etc. and wildlife and game management variables are points of interest as well.

The general file structure of the forest inventory is a flat file consisting of ASCII coded values. In the example shown in Figure 2 (which is already a simplification of the original data set for a given sample plot) the total number of lines to be used can be identified by the first two digits. For example, the first sample plot has a total of 25 lines. The first five are common plot information collected on the large fixed plot area of 300 m^2. The remaining lines indicate the number of recorded trees on the plot with their observed variables values.

It is obvious that such a system requires a great knowledge of the data itself and it is almost impossible to keep all the different variables in mind. The current situation of interacting with the system requires a forestry expert who, for many interpretations of the data, needs to be consulted. On the other hand researchers of other disciplines have got an increased interest to investigate forest areas because they represent our highest developed ecosystem and it is assumed that under changing environmental conditions (global warming, etc.) forest areas will respond very quickly and sensitively. This is because forest is considered as one of the key indicators for detecting and estimating environmental caused changes. It is important to note that one of the most complicated parts in quantifying the external impacts on our ecosystem is to distinguish between natural stress factors as they result from growing competition within a stand and the external impact caused by air pollution itself.

It can be summarized that we have a very well designed example for a forest inventory with a lot of information about one of our most complex ecosystems. On the other hand we have non forestry experts (ecologists, wildlife people, botanists, etc.) who have an increased interest in our forest data but access of the existing information is hardly possible for them. Another important reason which also suggests the urgent need of an adequate database reengineering is that the collection of environmental data is very expensive and whenever researcher have the chance to use already existing information a lot of money can be saved.

3 REENGINEERING PROCESS

For reengineering flat file structures we propose an approach consisting of four main building blocks and several major design activities. Two of the blocks are again decomposed into further development steps. Figure 1 gives an overall picture of the methodology in terms of life-cycle phases.

Building_block_1: Existing System
The first building block is the *existing system* and the activity to be performed is analysis and redocumentation. *Redocumentation* is focused on understanding the so-called mission of the existing system. The outcome of this phase is a set of requirements specifications, which are usually rather informal, disorganized, and expressed in natural language at this step of design. These specifications are used as a basis for *reverse engineering* of the existing system.

Building_block_2: Entity-Relationship Model
Reverse engineering is understood as the process of analyzing the existing system in order to identify its components and their interrelationships and to create a representation of the system in a high level of abstraction. The ER model is used for this purpose. This phase first consists of determining the entity types, their attributes and corresponding domains, as well as the integrity properties that must hold. When the structure of the entity types is determined candidate keys and foreign key groups must be determined. Candidate keys are used to uniquely identify each entity and foreign key groups help in determining relationships between entity types or to find generalizations. The development of an ER model is an iterative process, often involving trial and error. We refer to this process as *restructuring*. It is essential that the designer gets an exact understanding of how the different components of the ER model interact and what exactly their meaning is. In order to develop a proper solution experienced designers as well as a close interaction with application domain experts is necessary during this phase.

Building_block_3: Experimental prototype
This block is in close relationship with block_2 and consists of the development of an experimental prototype of the application in order to verify the design. With real examples

from the application domain potential users of the system may test its appropriateness and design errors at the conceptual level can already be found at this early step of design. This phase is performed by classical forward engineering and stepwise refinement. With *forward engineering* the process of moving from high-level abstractions and logical implementation-independent specifications (i.e. the ER model) to the physical implementation (the experimental prototype) is meant while *stepwise refinement* refers to the iteration process of refining the prototype based on corrections at the high-level specifications and vice versa. This process must be continued until the prototype implementation completely meets the users requirements.

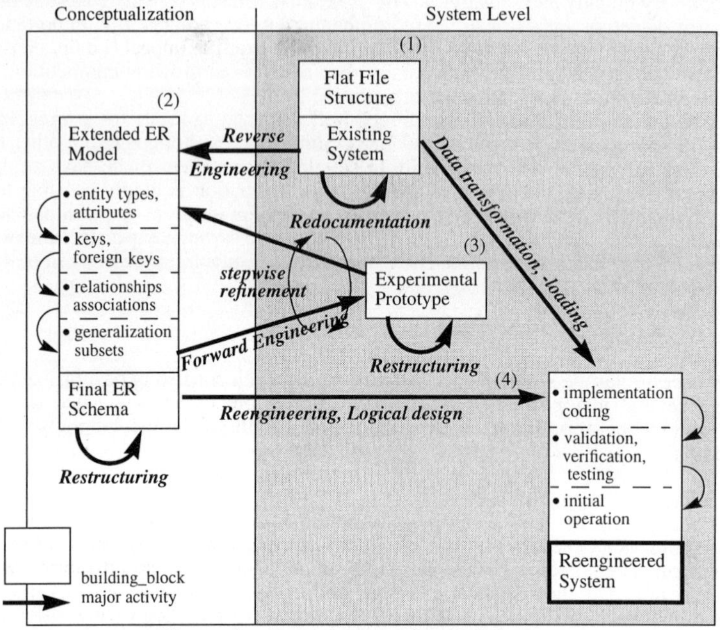

Figure 1 Reengineering Methodology

Building_block_4: Logical design, reengineered system
While the earlier phases of design are more intuitive and rely on the experience of the designer for this block it is possible to give some more exact rules how the new form of the system may be determined. Input to logical design is the final conceptualization of the application. *Logical design* consists of the transformation of the system independent high-level abstraction from block_2 into the representation of the data model of the reengineered form of the system, in our case in the relational data model. In the following we look at each concept of the ER model in detail to see how it can be transformed into concepts supported by the relational model:

Entity Types, Weak Entity Types and Multivalued Attributes
For each entity type E of the ER schema, we create a relation schema RS which includes all attributes that can have atomic attribute values only. By analyzing the functional dependencies valid in RS, the set of candidate keys may be derived, and one candidate key is chosen as a primary key of RS. Foreign key attributes or relationship attributes are not included yet; they will be added during subsequent steps.

For each weak entity type (WE) (i.e. entity type which may not exist without associated strong entity type) associated to E (strong entity type), we need to create an additional RS and include the attributes with atomic domain of WE in RS. In addition, we include the primary key attribute of the relation that resulted from the corresponding strong entity type E as a foreign key into RS. The primary key of the relation schema that resulted from WE is the combination of this foreign key and the partial key of the weak entity type.

For each multivalued or composite attribute (i.e. attribute having a non-atomic domain) A in E we create a new RS that includes an entity set X corresponding to A plus the primary key K of E.

Relationship Types

To transfer relationship types into a relational database schema we have to distinguish between <1:1>, <1:N>, and <N:M> cardinalities as well as between mandatory and optional participation of the corresponding entity types in the relationship type.

In general, two entity types E1 and E2 participating in a <1:1> relationship type result into two separate relation schemas. However, in the case of total (mandatory) participation of E1 and E2 we can also consider the possibility of merging them into a single RS.

For each binary relationship type of cardinality <1:N> we first have to identify the entity type E1 on the 'many' side. If the participation is mandatory, we include the primary key of relation schema RS2 that represents the entity type at the other side of the relationship type as a foreign key in RS1. If the participation of the entity type participating on the 'many' side is optional, we can also consider to set up three relation schemas; one for each participating entity type and one for the relationship type.

For each relationship type of cardinality <N:M> between two entity types E1 and E2 we develop an additional relation schema RS and include both primary keys from E1 and E2 as attributes. These attributes also form the primary key in RS.

Generalization, subset hierarchies

A generalization hierarchy of the ER model is transformed into a relation schema RS1 for the generic entity type and a separate relation schema RS2, ..., RSn for each of the specialization types. RS1 contains the generic entity type key and all common attributes including the common attribute on which the decomposition is based. Each specialization relation (RS2, ..., RSn) contains the generic entity type key as well as all attributes specific to the subsets. There is no difference in generalization and subset hierarchy in mapping into the relational data model, except interrelation integrity properties.

Logical design results in relational schemas which can be implemented by using the data definition language of a relational database management system (DBMS). In building_block_4 *implementation* and *coding* is concerned with the programming of the final operational version of the reengineered database application, the definition of the database schemas and views, the specification of the users and user groups and the definition of authorization and access control structures. *Validation*, *verification*, and *testing* are aimed at assuring that each previous phase of development is of acceptable quality and is an accurate transformation from the previous phase. The operational phase includes the actual data loading and an *initial operation phase* may be necessary to correct errors that were not detected in earlier phases. For data loading transformation programs must be provided that map the data of the flat files of the initial system into the representation of the reengineered system.

4 CASE STUDY

Based on an existing file structure a first ER diagram for the forest inventory has been reengineered. In this section we draw our attention on a specific subset of the data available from which we believe it might have potential for solving ecological questions. This subset is concerned with tree damages and related information.

In Figure 2 a general overview of the redesign process with respect to 'damages' is given. The process starts with identifying relevant information in the flat files, continues with developing an ER model (this step includes close interaction with the domain experts) and concludes with transforming the ER into relational data structures. The redesign is driven by the following motivation: We distinguish two different kinds of possible forms of damages: biotic and abiotic. *Biotic* are all damages caused by living individuals like human beings, animals, insects, viruses or wood destroying fungus. With *abiotic* we refer to all external environmental impacts to the ecosystem forest like wind, snow, frost, heat, etc. and all kind of damages caused by them. Abiotic and biotic damages are disjoint, i. e. a particular damage may only be assigned to one of the groups. However, both classes of damages have also properties in common. In particular, these may be sample point information, tree related information, each damage has a unique code, is in a certain development phase, makes bio/chem control necessary, the tree has a certain chance to recover from it, and so on. These common attributes indicate, that a generic entity type may be a proper solution to represent damages in a database and because of the disjointness condition of the specializations, modelling a generalization structure may be adequate. As can be seen in Figure 2 the distinction into biotic and abiotic damages is based on attribute 'type' and both specializations have different additional damage relevant attributes. For abiotic damages high-diameter ratio, rem. crown length, crown class while for biotic damages animal, part, gallery are relevant.

In a further hierarchy we decided to represent in the ER diagram the fact that we have different types of damages caused by biotic reasons. To distinguish between them we take the fact that the damage appears on a certain part of the tree. The idea is that damages caused by insects, animals, etc. normally are apparent on typical parts of the tree because of the nutrition preferences and living rhythm of the organism. From the view of a botanist, entomologist, wildlife manager, etc. the damage affected tree part is important for identifying the species, the importance of the damage and its influence to the ecosystem, the chance to recover and its relevance for further diseases. For example if a tree is bark peeled by deer the real problem is not the peeling itself but the remaining wound is often the entering point for insects and decays which influences the future vigor of a tree much more than the peeling itself.

Considering the subsets we divided a tree into root, leaves or needles (depending on the tree species), the bark, the crown and the stem. Because an ecological system is never stabile it is a matter of fact that some tree growth affecting diseases or damages may be located on several tree parts depending on the development status of the disease. Obviously, allocating the damage into a single class is sometimes not possible. This is the reason we decided to model specialization entity types as subsets with the effect that the assignment of a certain biotic damage to several specializations is possible. From an ecological point of view this option seems to be very important because organism often extend their living environment because of the lack of nutrition in the phase of the aggregation.

When the structure of the ER model is determined, the final step consists of translating the ER model into relational database schemas. In Figure 2 this step is only indicated by showing a single table 'damage' resulted from entity type 'damage' with some sample values. A complete transformation will include a single table for each of the entity types of the ER model.

Two further things are important to note: First, the ER model as given in Figure 2 represents only an extract of the universe of discourse because relationships to other concepts like tree information, plot information, soil types, etc. are not explicitly given. Second, the ER is developed on an individual tree basis and thus is reflecting the structure of the forest inventory. With this we mean that an entity of type damage refers to a certain damage a particular instance of tree suffers from. This has the effect, that in a relational database a particular entry in a table 'damage' is uniquely determined by combining the sample plot identification with the tree identification (within the plot) and with damage identification. Including the damage code into the key is necessary because a single tree may suffer from several damages (as it is the case with tree 21 on plot 8039 in the example given in Figure 2).

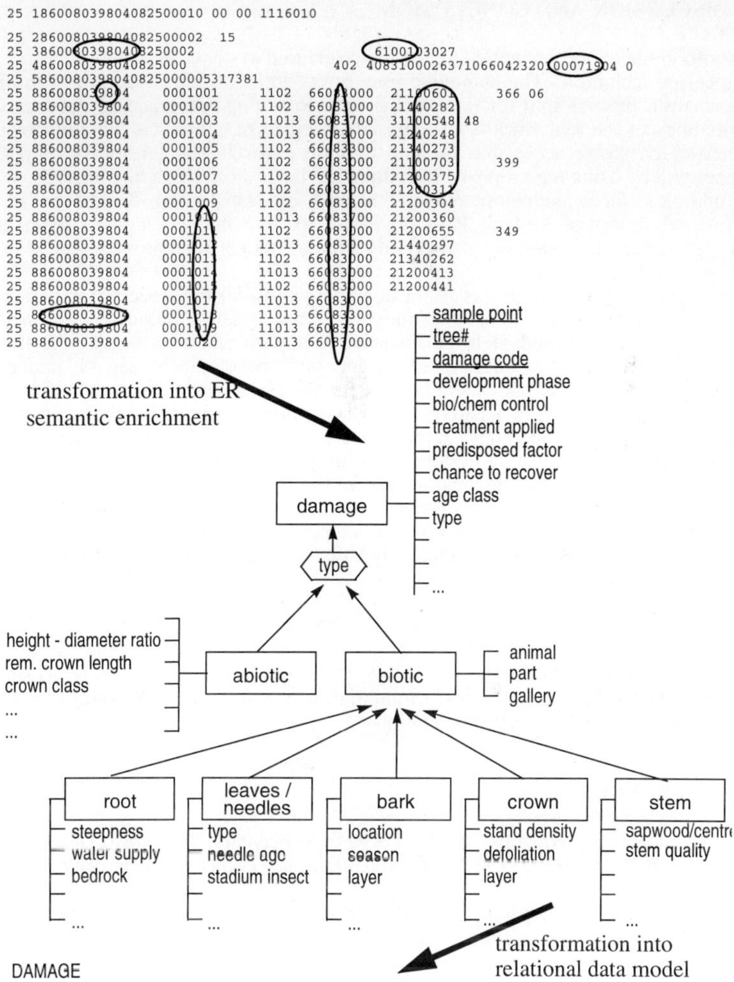

DAMAGE

<u>sample point</u>	<u>damage code</u>	<u>tree number</u>	development phase	chance to recover	age class	...
008039	17	21	beginning	high	60-81	...
008039	21	21	advanced	minimal	40-61	...
001232	07	12	medium	minimal	1-21	...
...	...	...	...	...	...	...

Figure 2 Case Study

5 DISCUSSION AND CONCLUSION

In order to enhance the semantics of flat file structured systems we have proposed a database reengineering technique. The technique uses Entity-Relationship modelling combined with a transformation process into relational databases. Resulting reengineered databases represent the semantics of the real world in a more advanced form which has the advantage that data becomes much easier accessible for non domain experts. The feasibility of performing a reengineering by using the proposed technique has been shown by a first analysis of one of the most complex forest inventories available and by transferring it into an reengineered experimental prototype system. It is important to note that our system represents a first prototype version of an ongoing project only and has not yet been proven as working properly by a large number of users.

The analysis described in this paper does only consider the static properties of a system. What is of interest is also to include functional and dynamic processes within the reverse engineering. This has turned out to be of particular importance in performing the case study. To be more precise, biotic damages/diseases and their development can be understood as a dynamic system. Starting from latent conditions, which normally do not yet effect the system negatively and often have a positive impact as a nutrition base for the useful animals the meaning of the disease can change rapidly and as a worst consequence it can even result into devastated forest areas. From a system modelling point of view it is interesting to look into modelling the conditions that are responsible for the change of the state of a disease in terms of object states and pre- and post conditions. Additional work has been devoted to this aspect in which data flow diagraming is used to model the functional part of a system while object behavior diagrams to capture the objects dynamics. First results of our efforts are reported in Pernul and Hasenauer (1995).

Acknowledgment
We would like to thank our students I. Haslik, G. Hentschel, G. Retter, R. Rosenthaler, M. Rupp, and M. Scheschy-Lun for performing part of the analysis of the forestry inventory and for producing the prototype implementation.

6 REFERENCES

Batini, C., Ceri, S., Navathe, S. B. (1992). *Conceptual Database Design. An Entity-Relationship Approach.* The Benjamin/Cummings Publishing Company, Redwood City, CA, 1992.

CACM (1994). Reverse Engineering. Special Issue of *Comm. of the ACM*, Vol. 37, No. 5, May 1994.

Chen, P. P. (1976). The Entity Relationship Model: Towards a Unified View of Data. *ACM Trans. on Database Systems (TODS)*, Vol. 1(1).

Markowitz, V. M., Soshani, A. (1989). On the correctness of representing extended ER structures in the relational model. *Proc. of the ACM SIGMOD Conf.*

Markowitz, V. M., Makovsky, J. A. (1990). Identifying extended ER object structures in relational schemas. *IEEE Trans. on Softw. Eng.*, Vol. 16, No. 8. Aug. 1990.

Navathe, S. B., Awong, A. M. (1987). Abstracting relational and hierarchical data with a semantic data model. *Proc. of 6th Conf. on the Entity-Relationship Approach.*

Navathe, S. B., Pernul, G. (1992). *Conceptual and Logical Design of Relational Databases.* In: 'Advances in Computers', Vol 35, M. C. Yovits, (Ed.), Academic Press 1992, pp. 1-80.

OEFI (1994). F*orstliche Bundesversuchsanstalt, Instruktionen für die Feldarbeit der Österreichischen Waldinventur 1992 - 1996.* Fassung 1994. Eigenverlag der FBVA.

Pernul, G., Hasenauer, H. (1995). Combining reverse with classical forward database engineering - an object-oriented approch to solve the legacy system dilemma. *Proc. 6th*

Int. Conference on Database and Expert Systems Applications (DEXA'95). London, UK, Sept. 1995. Springer Verlag (LNCS), 1995.

Premerlani, W. J., Blaha, M. R. (1994). An Approach for Reverse Engineering of Relational Databases. In CACM (1994).

Software (1990). Maintenance, Reverse Engineering and Design Recovery. Special Issue of *IEEE Software*, Vol. 7, No. 1. Jan. 1990.

Software (1994). Systematic Reuse. Special Issue of *IEEE Software*, Vol. 11, No. 5. Sept. 1994.

Teorey, T. J., Yang, D., Fry J. P. (1986). A Logical Design Methodology for Relational Databases using the Extended ER Model. *ACM Computing Surveys*, Vol. 18.

7 BIOGRAPHY

Univ.-Ass. Hubert Hasenauer received MSc and the PhD degrees from the University of Bodenkultur, Vienna, Austria, where he currently holds a position as a full time researcher and lecturer. Prior he was a visiting scientist at the College of Forestry and Wildlife Resources at the Virginia Institute of Technology, Blacksburg, VA. His major research interests are inventory design, growth simulation, and computer applications in forestry.

Univ.-Doz. Günther Pernul received MSc and the PhD degrees from the University of Vienna where he currently holds a senior lecturer position. Shortly he will take over a position as a full professor of information systems at the University of Essen, Germany. His main research interests are information systems analysis, modelling and design, computer security, and database management systems.

7

Complex Metadata management in Earth Observation for environmental research

M. Zingler, H. Pintaritsch
European Space Agency - ESRIN
Via Galileo Galilei, I-00044 Frascati, Italy, Tel. +39.6.94180.1,
mazi@mail.esrin.esa.it

Abstract

User expectations and requirements for access to Earth Observation data and information are constantly increasing. Recent improvements in open system architectures and networks have intensified discussions on interoperability and data availability. Efficient Metadata management software supports the search for data and the final retrieval from EO archives. The European Space Agency is challenged to provide solutions for that objective. Environmental research applications will benefit from enhanced metadata layer functionality to access EO products.

Keywords

EO metadata, inventory system, provider/subscriber concept, thematic access

1 INTRODUCTION

Earth Observation (EO) is the combined effort of technical and scientific resources to achieve a variety of goals e.g. understanding of large scale phenomena on earth (Global Change), generation of maps and recording of atmospheric processes for meteorology. A large part of the EO user services manages metadata (MD) with the objective to lead users to the products they need in their application domain. MD should present a summery on each required data set. MD systems can assist the user during the process of mining into data collections he is interested in. The MD models might also be exchanged between systems for interoperability purposes.

The European Space Agency takes a major share in the EO framework through its remote sensing satellites and the operations of ground segment (G/S) facilities. The EO G/S should foster the completeness, integrity and accessibility of the data obtained from the instruments.

The capability of accessing even small portions of data with short response times has to be ensured in spite of a rapidly increasing volume of stored data (up to Tera- or Pentabytes).

The proposed data management principles discussed below focus on operations on MD as required by satellite data inventory systems.

2 THE EARTH OBSERVATION USER SERVICES FRAMEWORK

Earth Observation covers large geographical areas, but records as well small scale processes and features. This is a result of the increasing precision and resolution of the instruments. By its nature EO crosses boundaries and thereby becomes an international issue. EO has to consider the interests of local groups with a large geographical distribution. The broad variety of scientific and technical applications with their demand to perform cross-disciplinary data mining (Thieman 1992) is relevant to EO system design.

The requirements arising from the mission to serve those communities pose complex architectural design tasks on the agencies. Our target user communities are located in the scientific community (research institutes, universities, etc.), institutional community (meteorological offices, governmental institutions, etc.) and the commercial community (value-adding services, cartographic services, environmental services, etc.). The European Space Agency undertakes efforts in collaboration with the EC to create an European Earth Observation System that will cope with user group and system heterogeneity. NASA's EOSDIS can be named as such a system, although being adjusted to the scientific community (Elkington 1995).

Complex requirements stem originally from the scientific community. Now commercial interests play more and more a substantial role in the advertisement and distribution of EO data, and some industrial activities depend on high quality data with high reliability in distribution (e.g. regular Fast Delivery Products for ship route planning). Figure 1 shows the modular character of the overall ESA EO data and user services capable of:

- Advertising data acquisition possibilities,
- Advertising products, services and costs,
- Providing user tools and system access gateways,
- Providing service navigational tools, Directories, Guides and tutorials,
- Providing inventory and browsing facilities (including CD-Browsers),
- Providing Data Dissemination facilities,
- Providing user registration, access control and billing,
- Providing Subscription services of functions described above.

An EO data and service site assembles several of the above components. They have to cooperate at and between distributed sites, which requires interoperability layers that act as gateways from/to client systems to the various server systems. Depending on the type of services the interoperability layer will be achieved through concepts from common protocol for information request and exchange up to direct access to resources at the sites. The user system domain is coupled via firewalls to a dedicated operations system domain and servers. A complex cooperative hypertext Guide and Directory systems forms part of the advertiser and navigation role of the system.

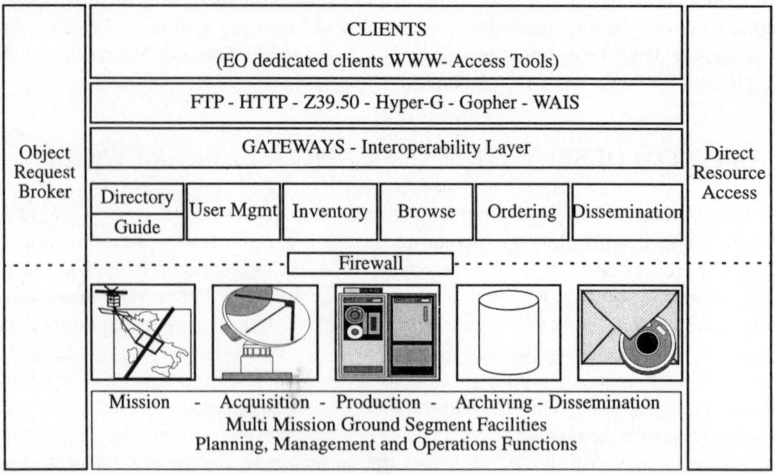

Figure 1: EO User Services Conceptual Model

The system design has to anticipate a constant growth of data, notably when automated classification of EO products becomes more common and adds attributes and relationships to the model (Cromp et al., 1992). An adequate data architecture in the EO system (Childs 1991) and a consistent modelling of EO (meta-)data are the basis for high-level interoperability. As an additional aspect, user server systems have to provide flexible service facilities (e.g WWW, Object Broking), while the operational side requires robustness and stability. The latter has to maintain the availability and integrity of the satellite data for 15 to 50 years.

3 A METADATA CONCEPT IN EARTH OBSERVATION

A focal point for users of EO information is the knowledge about existing data sets (collection of granules) including related auxiliary data as obtained e.g. from ground campaigns. The granules from that data are stored in distributed archives, where only partial coherency of granules is achieved since the conditions of data reception and post-processing are varying from center to center. Granules are the smallest units of an identifiable data sequence, e.g an image.

Metadata describing each granule, called Metagranule (MG), can be created at the planning stage of a granule (for future satellite acquisition) or upon registration. Their lifecycle is determined by the status of the granules they relate to. The MG makes particular sense through the object-oriented modelling by which the rigid record and table driven model is broken up and the underlying granule is described more accurately.

In general, the metadata (MD) describe the physical and logical properties of the stored data sets and consist of textual information as well as low resolution browse images or statistical summaries of granule properties. The attributes in a metagranule are determined by various sources (user, telemetry, production etc.). MGs might be aggregated up to an acceptable temporal granularity (Hood et al. 1991). MD allow users to judge if data sets are suitable for their application and form an important element in the chain from user's first contact with an EO advertising service to his final order of a product. Browse products (low resolution products) and directory information (i.e. a hierarchical representation of available products) can also be regarded as metadata and will be logically co-located and accessible with the textual inventory information for a product.

A data pyramid reflects the data model that must be supported with each level having an index to the lower level in the pyramid, showing the increasing accessibility to the top and the large volume of products at the bottom of the pyramid. The low volume of condensed directory information is the located at the top. We extended the pyramid representation as proposed by (EOSDIS 1994) to a data rhombus to show the related service function and its properties with low (0) and high (∞) requirements. The embodied data mining process is the summery of the user's activities to obtain the target product and is discussed later.

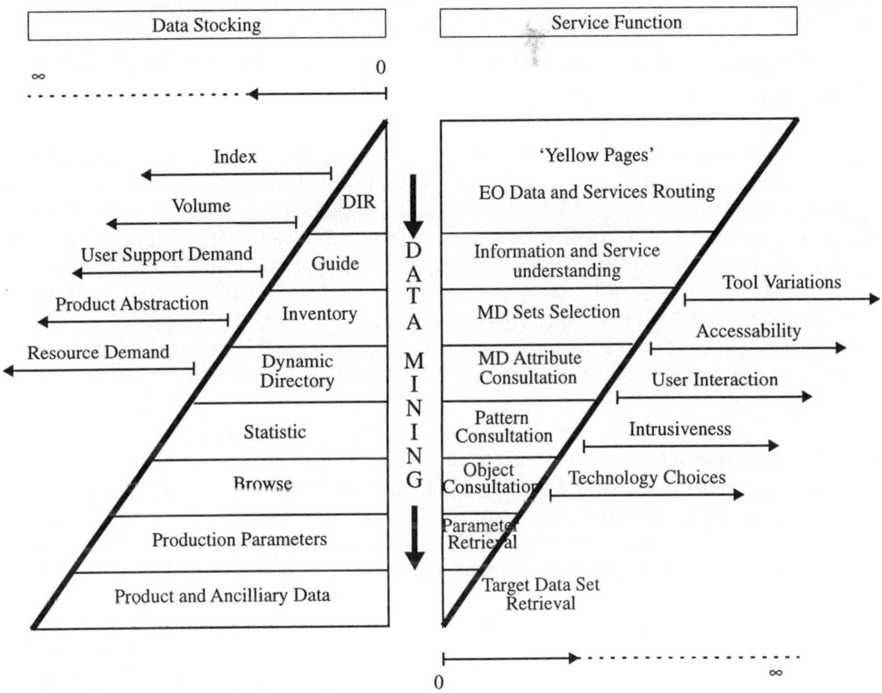

Figure 2: Data and Service Function Rhombus

The distribution of providers and users and the dispersed service nodes enforce global and local metadata models. The system analysis process had to find mechanisms to ensure that the system could cope with the parameter variation in the model. A major role plays the possibility to assign to each MG type particular interpretation, processing and referencing methods as they are required by each user group. These are best implemented as methods on object data. MMIS (see below) provides an integrated schema evolution mechanism for the complex MD structure. It is written in an IDL-like fashion and called Universal Inventory Definition Language (UIDL). User views, views by the updating entities and the internal OODBMS schemas are defined in a single description process and are compiled into an Interface Definition Object. This can be communicated directly to other federated inventories of the same kernel type. It supports therefore ideally the Provider/Subscriber concept and could contribute to the implementation of a more global information resource dictionary as described in (Hsu et al., 1991).

We consider the following five mechanisms as relevant to metadata management:

- the provider-subscriber concept for a controlled data replication to distributed systems.
- the thematic access to improve access and performance in inventories.
- the object oriented modelling.
- data mining support.
- the UIDL to communicate the three-folded data structure.

4 METADATA MANAGEMENT IN THE MULTI MISSION INVENTORY SERVICE SYSTEM (MMIS)

Increasingly, the object-oriented paradigm is applied in payload data systems design, metadata modelling and management. Major constituents in related applications are object-oriented modelling techniques, Object-Oriented Database Management Systems and Object-Request Broker technologies. The *Multi Mission Inventory Service System* (MMIS, Figure 3) is a development managed by ESA/ESRIN demonstrating new concepts of metadata management based on the above technologies. It is one system component of the *Multi Mission User Information Services* (MUIS) which is based on a conceptual architecture as shown in Figure 1.

Metadata, generated or available at distributed sites, is kept in the inventory which is established as an object-oriented distributed database (Öszu 1991). The main EO inventory application qualities are:

- The DB application is permanently loaded by a stream of updates from the various sources (high operational transactions). The share of Near-Real-Time updates will be a factor for the future (NRT product availability and management).
- The actual user transactions (queries) load is still low, but will dramatically rise when the system is on-line over the WWW on Internet (expected are 500 to 10.000 queries daily).
- The distributed function across the provider/subscriber links will impact performance requirements, since the replication mechanism must access consolidated MD in the provider DB.

The task in the federated inventory network is to allow access to global and local metadata schemas and to pass context information in case of distributed searches and service requests during user sessions to other sources of information. An object and service broker will dispatch requests and returned information. The interoperability on the advertiser level will be achieved over HTTP with some extensions to allow the invocation of sophisticated server functionality which then uses e.g. DCE or CORBA.

The logical MD management architecture is a mixture of a cooperative C/S (Client/Server) architecture and federated DB systems. The C/S cooperation results from the way how end-

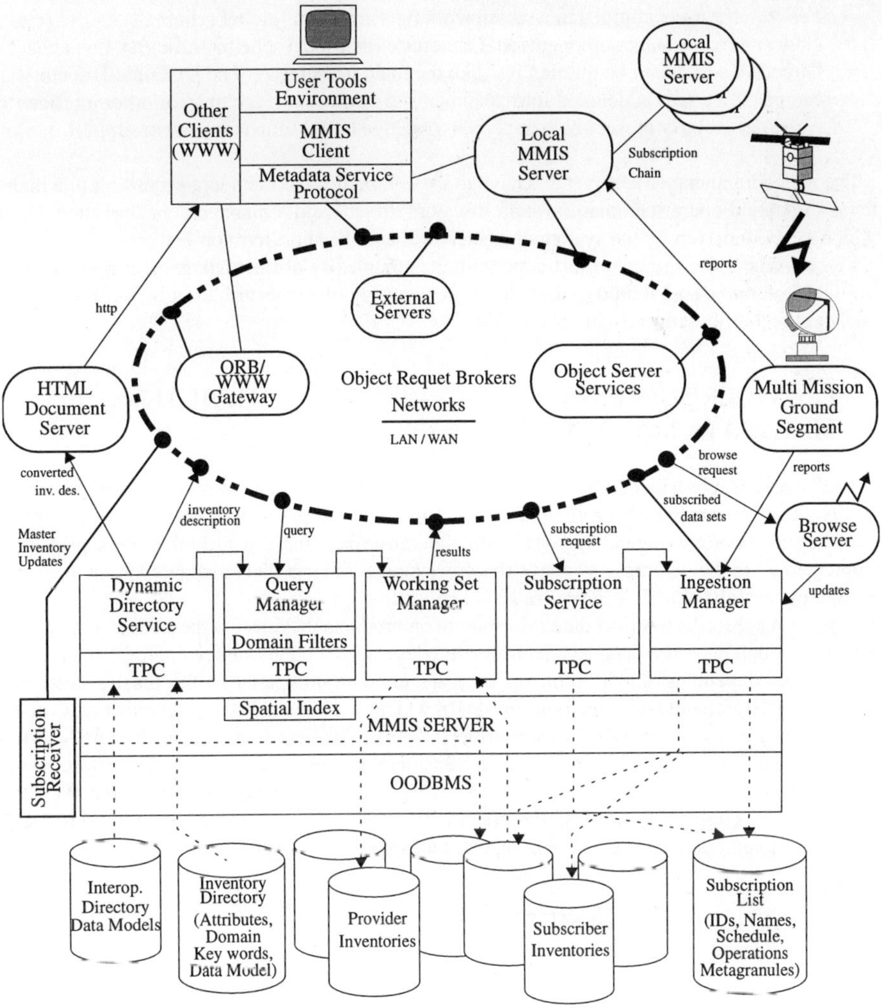

Figure 3: MMIS Services and Conceptual Architecture

user tools interact locally and over large geographical distances. Another factor is the user's heterogeneous environments with discrepancies in connectivity and performance. The MMIS protocol considers restrictions on the client side as it allows server oriented operations and pulling of results in chunks from the client working set. The main concepts for distribution are the:

- *Federated system* components which work on a common 'global schema'.
- *Cooperative system* components that announce their local schema through a Dynamic Directory which can be queried just like the main inventories. The Dynamic Directory can be converted on demand into documents to allow text searching for other interface types particularly those which apply a knowledge based semantic system support.

The metadata management systems have to survey the flow of metadata according to a high-level strategical operations plan (system initiators driven) and a management operations plan which are both driven by the system characteristics and the short term objectives.

The metadata management must cope with the complexity of the management rules including fault tolerance and should guarantee the consistency of the information between a central site and the distributed sites (Zingler 1995).

5 PROVIDER/SUBSCRIBER CONCEPT FOR A DISTRIBUTED METADATA MANAGEMENT

The Provider/Subscriber (P/S) concept is based on the notion that other sites want to or must receive a set of metadata on a regular bases. The subscription is important since it allows to establish a persistent logical link between the subscriber and the provider. This principle, implemented in the inventory system software, supports the metadata management in an architecture of centralized and decentralized EO services.

Users can subscribe to up-to-date information on products they need. Operations as replicate, update and delete on the data sets determine the level of data consistency amongst the distributed copies depending on the moment they are initiated compared to the master inventory update. A P/S protocol had to be built in MMIS which maintains a list of subscribers and operations that are requested by those subscribers. The mechanism is based on the ORB with a scheduler and can enforce operations also after network link interruptions. The subscription can be chained to give remote sites the possibility to act in turn also as a provider. We have considered this approach as fitting the operational needs of inventory servers. Nevertheless it can be complemented later by a client driven status update as described in (Joy et al. 1992).

6 MODELLING OF EO METADATA

The modelling of EO metadata has to cope with a variety of instrument and data types and their properties in terms of:

- Frequencies and techniques (spectral ranges, multi-channel devices, active, passive),
- Coverage (temporal and spatial scale, orbit characteristic, repetition or permanent coverage, instrument footprint),
- Resolution (spatial, temporal, radiometric, spectral resolution),
- Instrument geometry (nadir- and limb-looking, side-looking, view-angles),
- Calibration (information about stability, device and data calibration),
- Data rates and observation of events (e.g. volcano eruption process).

There is large amount of different instruments for different purposes. The main classes are Spectrometers (spectral), Radiometers and Scatterometers (intensity), Imagers, Sounders and Altimeters (spatial) (Kramer 1992). Hence, dozens of product parameters exist and the variation in parameter characteristics is so large, that its impossible to take them into account a-priori a system design (Hachem et al. 1990). Especially the geometry (footprints) of the instruments is noteworthy, since it is special challenge to databases that have to keep a knowledge of the spatial reference of each product.

The management of EO data is confronted with the challenge to handle complex data structures. Products of the most advanced instruments contain in addition to temporal, spatial and instrument specific information, high level quality and processing parameters. Experienced users want to create queries that include a combination of those parameters, and sometimes even post-processing algorithms on them (e.g. restricted view angle of instruments).

This aspect leads to the creation of a data model and database indices which are based on previous experiences of average user behavior. Nevertheless they have to be flexible enough to be extended when new user communities set up new priorities.

The high importance of spatial information in metagranules led to the use of R*trees as the basic indexing method (Guttman 1984, Beckmann et al. 1990). The nodes in the trees are minimum bounding rectangles (MBRs) of arbitrary shapes, e.g. polygons. This indexing technique suits the use of virtual metagranules which can consist of references to several real metagranules. The combined information is transparent to the user, who can search on them just as on usual data.

7 DATA MINING AND THEMATIC ACCESS TO EO METADATA

The data access activities of users are characterized by an interest in a specific application area that he or she is working in. It is important to the acceptance of the system by the user that required information is available at the right time. Hence, the system analysis process has to consider 1) the *EO User Access Model*, defined by specific logical operators and reflecting typical user behaviour, and 2) the *EO System Metadata Model*, describing data structures and properties and stating the physical availability of metadata that is distributed over different repository sites.

The data mining problem described here comprises the search from top-level interactive (searchable) guides and directories, inventories, browsers and order services to a low-level archive oriented product query and/or retrieval mechanisms. Those queries and commands can be issued by the user and also by the system itself. An 'advertise' service, available at the level of the guide subsystem, supports the entry to the system and is linked to the metadata. The further data mining process follows the logic in Figure 2, the data rhombus.

A solution to multi-discipline query processing is to provide a metadata retrieval mechanism that considers the variety of thematic query combinations. The search engine has to recognize that the query matches a thematic pattern and traverses the index tree. It is made of two layers:

- the *thematic layer* organizing application typical queries into an index structure.
- the *spatial and temporal layer* reflecting the common reference properties of MGs.

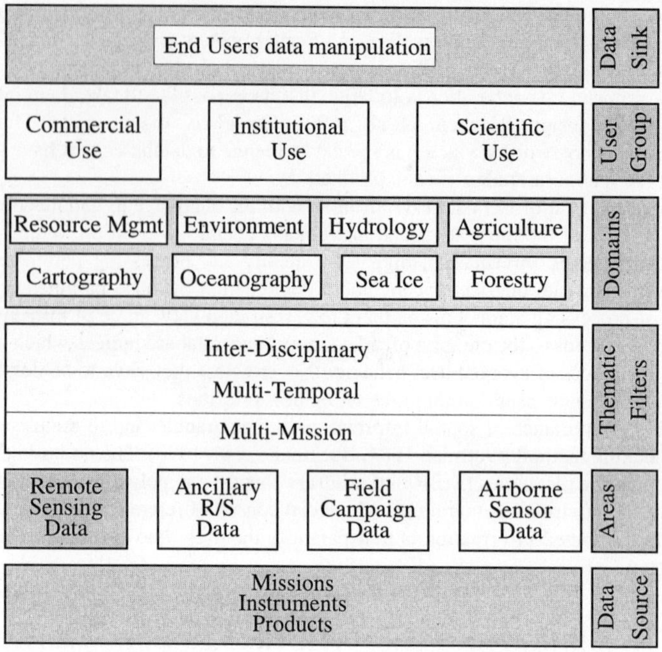

Figure 4: EO data layers

This mechanism known as thematic access is a fundamental approach for efficient searches in large scientific metadata repositories (See Figure 4). It can replace or complement ad hoc object-oriented query languages. Another important aspect is the availability of metadata to users. This is achieved either simply by conveniently putting metadata onto a local server, or also by the temporal requirements of certain application for regular updates. The availability of metadata is ensured from the user perspective by:

- subscription to a particular set of metagranules specifying the thematic and other properties supplied by a provider,
- directory services announcing hierarchically the existence of data sets (yellow pages).

In the MMIS data is arranged and partitioned in advance, creating domains. This significantly increases the precision of search results and reduces the response time. Domains will naturally be created for spatial and temporal criteria, and lead therefore to the common use of abbreviated search parameters (e.g. the user can directly search for the region 'Italy' instead of approximating its borders with a polygon). This differs from the approach of synonym lists.

As OODBMS allow several entry points into a database, every domain can immediately be accessed. Additionally, the partitioning of domains can be done recursively, making domains composed of several subdomains. This method is used to improve the database 'layout' and significantly reduces search times.

In the OO environment the domain concept can very easily be applied to *all* metagranule parameters, i.e. grouping can be done on qualification parameters. Using this technique, data from different satellite missions but with similar parameters can e.g. be grouped to form a domain of 'Oceanography' to see the effects of environmental change over several years, seen by various instruments and satellites

8 FUTURE IMPROVEMENTS TO EO METADATA SERVICES

The move from a canonical query form to real object-oriented queries is a step that is induced where metadata with a complex model has to be searched and also where location transparency should be achieved through the Object Request Broker mechanism. There are practical considerations that derive first an object oriented ad hoc query language (OQL) whereas the OO query object approach can increase the performance and reliability in a distributed system.

The system evolution has to provide search and management mechanisms that guarantee in the future a low-level access directly to archived granules, using the index system mentioned above with metagranules pointing directly to the archive. Since the complexity of that search can be considerably high an invocation of *search-agents* launched at the user discretion is proposed. The agent would interface with the storage management system, complete a search matching the criteria and react on temporal constraints at the host system. The agent-driven approach search increases the abstraction of data location and allows to return different data representations besides other functionality that can be off-loaded from the local user client.

An important step forward in the area thematic access will be the integration of a knowledge-based methods that optimize and expand the thematic index substructure of the inventory.

9 CONCLUSION

The management services of actual and future EO systems follow generally the tradition of distributed information systems, but add more complexity to it through the implementation of a highly flexible distribution. This is and will be enforced by local, regional and global data location and distribution requirements. The metadata handling is an important element in the EO system design because it describes what kind of data sets the system preserves in its storage. The user must able to reach through the metadata level the required data set.

The solutions found for the management of EO metadata in a highly distributed environment and the design of ESA/ESRIN's Multi Mission User Information Services must provide the needed facilities. As demonstrated before, some solutions should be extensible to other similar scientific and environmental applications.

10 REFERENCES

Beckmann N. et al. (1990) The R*tree: An Efficient and Robust Access Method for Points and Rectangles. Proc. of ACM SIGMOD Int. Conference on Management of Data, 322 - 331.

Childs D.B., (1991) Data Architecture and the Data Architect, in *Earth and Space Science Information Systems* (ed. A. Zygielbaum), Conf. Proc., 50-564.

Cromp R.F., Campbell W.J., Short N.M. (1992) An Intelligent Information Fusion System for Handling the Archiving and querying of terabyte-sized spatial Databases, in *Earth and Space Science Information Systems* (ed. A. Zygielbaum), Conf. Proc., 586-597.

Elkington M.D. (1995) EEOS Concept Study - EOSDIS Architecture Review, ESA/ESRIN Contract P/WS/KH/9681.

EOSDIS (1994) System Design Specification for the ECS Project, Document Number: 207-CD-001-001, 6/94.

Guttman A. (1984) R-trees: a dynamic index structure for spatial searching. Proc. of ACM SIGMOD Int. Conference on Management of Data, 47 - 57.

Hachem N.I., Gennert M.A., Ward, M.O. (1990) A DBMS Architecture for Global Change Research, in *Earth and Space Science Information Systems* (ed. A. Zygielbaum), Conf. Proc., 121-131

Hood C.A., Howie R., Verhanovitz R. (1991) The development of a Prototype Level-Three Interoperable Catalogue System, in *Earth and Space Science Information Systems* (ed. A. Zygielbaum), Conf. Proc, 152-165

Hsu C. (1991) Information Resources Management in Heterogeneous, Distributed Environments: A Metadata Approach, IEEE Transact. on S/W Eng., Vol. 17, No. 6, 6/91.

Joy S.P., King T.A., Walker R.J. (1992) Automating Database Management for Distributed Database Systems, in *Earth and Space Science Information Systems* (ed. A. Zygielbaum), Conf. Proc, 444-453.

Kramer H. J. (1992) Earth Observation Remote Sensing - Survey of missions and sensors, Springer Verlag, 200 ff.

Özsu M.T., Valduriez P. (1991) Principles of Distributed Database Systems, Prentice-Hall, 66 ff., 494 ff.

Thieman J.R. (1992) Identifying, Accessing, and Interoperability using Earth and Space Science Information Systems, in *Earth and Space Science Information Systems* (ed. A. Zygielbaum), Conf. Proc., 871 - 883

Zingler, M. (1995) Multi Mission Inventory System Development Guideline, Technical Note, ESA/ESRIN, MMIS-DGD-TN/2.0

11 BIOGRAPHY

Matthias Zingler received a MSc degree as an engineer for Aerospace and Space Technology from the University of the Armed Forces, Munich, 1986. He worked in the area of missile air-defence systems and joined the Nato Programming Center, Belgium, in 1989 for developments in missile command and control system software. Since 1993, being with the European Space Agency as a Technical Officer in Italy, he manages component design and development in satellite payload data ground systems.

Hannes Pintaritsch received a MSc as an engineer for applied mathematics and informatics at Graz University of Technology, Austria, 1993. His thesis was partly funded by Hewlett-Packard as part of a 2 years information retrieval project. He joined the European Space Agency in 1994 as young graduate software engineer for object-oriented systems and databases.

8

The Ecological Information-Managing System of Bashkortostan Republic

R.Z. Khamitov [1], V.E. Gvozdev [2], S.V. Pavlov [2]
[1] Environment Protection Ministry of Bashkortostan Republic
[2] Institute for Problems of Applied Ecology and Natural Resources Use of Environment Protection Ministry,
12/1, 8 Marta Str., Ufa, 450005, Bashkortostan, Russia.
Telephone: (3472) 287779, fax: (3472) 283913,
e-mail: root @ ippeprb.bashkiria.su

Abstract

Nowadays the ecological situation in Bashkortostan Republic (BR) is highly strain. There is a powerful petroleum-chemical industrial incorporation which negatively impacts the environment condition and population health.
The decision of ecological problems of BR requires the systematic approach. It is necessary to develop the strategy and provision of ecological safetness of Republic.
The Institute for Problems of Applied Ecology and Natural Resources Use of BR is the leading organization in creation of the Ecological Information-Managing System of BR (EIMS BR). The main aim in development of this system is the realible data supply for Regional and Local Management in taking decisions on control, stabilization and improvement of regional ecological situation.
In these papers some approaches to EIMS BR creation and some practical results, obtained in EIMS, are described.

Keywords

Information-managing system, geoinformational models, databases.

1 INTRODUCTION

The Republic of Bashkortostan is an area of ecological disaster. In the first place it is connected with the fact that there are four powerful oil-refining plants operating in the Republic with the total working capacity of 55 mln tons of oil per year. 5O% out of these plants were equipped in the 5O's, 30% - in the 6O's and only 2O% - in the 7O's. No reconstruction of the enterprises has been carried out lately.

A wide range of chemicals such fertilisers, pesticides and herbicides, epoxy resins, sulfuric and hydrochloric acids etc., in all more than 200 chemical nomenclature titles, is produced at about 10 chemical plants located throughout the Republic. The equipment of the plants was basically installed in the 5O's-6O's. On account of outdated and imperfect production processes they are discharging industrial waste into the water and polluting the air with toxic substances, dioxins and bipheryls being the most harmful and dangerous among them.

In the Republic there were rich explored and blocked-out oil reserves. At present about 8O% of them are practically depleted. About 20 mln tons of oil are produced annually. Oil-producing technology does not comply with the modern standards and this results in pollution of vast of underground water-bearing horizons. Drinking water in many regions of the Republic is unfit for consumption.

Machine-manufacturing and instrument-building enterprises pollute the environment with heavy metals compounds as galvanic production industrial waste are discharged into the city sewerage without any preliminary purification.

The basic fuel for operating of power plants - electric and thermal power stations - is natural gas. In rare cases it may be fuel oil or even more seldom coal may be used. However the masters of these plants are discharged into the atmosphere without being subjected to any cleaning at all, the volume of discharge running into 150,000 tons per year.

Today one of the main problems is the air pollution with exhaust gases from automobiles since there are no catalytic converters in the exhaust system. In many cities ethyl lead-doped petrol is used.

Deposits of zinc, copper and metallic ores are being exploited in the Republic. The production upgrading processes being rather imperfect, sulfur, heavy metals, arsenic and other compounds are being gradually accumulated in waste dumps. This isresulting in harmful damage and disruption of environment.

About 2 mln tons of detrimental substances are annually discharged into the air, the population of the Republic constituting 4 mln people. About 1.2 billion cubic meters of industrial waste and domestic refuse are discharged into the water and only 10% of the waste water amount is purified. In territory of the Republic there are accumulated about 600 mln tons of industrial waste, about 15 mln tons of which can be referred to as toxic ones. These waste are not processes and thus are badly damaging the environment.

In the Republic there are no plants for treatment and utilisation of domestic refuse. They are brought to and stored on most primitive proving grounds supplied with no systems or discharged water drainage, gas gathering and refuse grading.

Large-scale pollution of the environment in the Republic caused the water supply contamination in Ufa. Dioxins compounds which are strong poisons were found in drinking water. In 1990 drinking water was contaminated with phenol. There is an extensive growth of cardio-vascular and oncology diseases in urban and areas in the Republic.

The number of children suffering from congenital abnormalities has substantially increased. The majority of the people are experiencing a heavy psychological stress caused by the factor of the environmental contamination and by existing in the conditions of a constant pressure of ecological disaster.

2 THE ECOLOGICAL INFORMATIONAL-MANAGING SYSTEM OF BASHKORTOSTAN REPUBLIC

The decision of ecological problems of Bashkortostan Republic requires the systematic approach; it is necessary to develop the strategy and provision of ecological safetness of Republic.

To fulfill the main trends of mutual activity the information exchange and providing with the assistance is suggested in solving the following problems:

1) creating and practical application of the ecological information system for:

- obtaining, acquisition, systematisation and long-term keeping of environment information
- arrangement of ecological information exchanging between different organisations belonging to different fields of activity; informational supporting of decisions aimed to environment improvement
- the objective estimation of existing ecological conditions, short-term and long-term predictions of ecological situations
- risk assessment of extraordinary situations occurrence, development of measures on their preventing and informational support of decisions aimed to eliminate in case of occurrence.

The Institute for Problems of Applied Ecology and Natural Resources Use (IPAENRU) of Bashkortostan Republic which was founded in December of 1993 is the leading organization in creation of the Ecological Information-Management System of Bashkortostan Republic (EIMS BR).

The main purpose of creation of EIMS BR is the effective ensuring with reliable information for State and local authorities, for different enterprises to adopt managing solutions and to coordinate the efforts on management of ecological situations in the region, to arrange the rational usage of natural resources, to prevent the availability of extreme ecological situations occurrence.

The EIMS may be distinguished by the following components (functional subsystems):

- subsystem of acquisition and primary processing of ecological data
- subsystem of data transference
- subsystem of storage and data systematization
- subsystem of processing and submitting of results of data processing.

Functional scheme of EIMS BR is shown on Fig. 1, where S1,...Sn - are the sources of ecological information, U1,...Um - information users.

The subsystem of acquisition and primary data processing is designed to obtain the information from different territorial and department located sources of data about of environment quality and the sources of pollution.

The subsystem of data transference provides with technical availability of information exchange between different enterprises and directing bodies. The telephone and telegraph channels of common communications are applied on the first stage of EIMS creation during the maintaining of basic functions and principles of ecological information usage. The effective data exchage to adopt extreme decisions is supposed to be carried out by special channels of communication.

Receiving of the information from almost inaccessible regions is supposed to be done by the application of satellite ways of communication which are highly developing in BR nowadays.

The subsystem of storage and systematization of ecological data is designed to arrange the longtime storage of information on computer carries. This subsystem presents the distributed data base with maintenance by special body - data base administrator. Logical data base is intended to aggregate local data base of enterprises into the unique distributed data base. The scheme of arrangement of data exchange in the distributed data base is given on Fig. 2.
The subsystem of processing and preventing is designed for informational supportance of directing bodies decisions and coordination of actions directed on ecological situation management in the region. This subsystem is to realize the following functions:
- assessment of current and forecasting of ecological conditions changing
- intellectual interface of user to obtain the information stored in distributed data base
- simulation and estimation of consequences from taken decisions in the sphere of ecology and natural resources use
- information presentation including spaced one with GIS usage.

The structure of subsystem of processing and preventing of the information is shown on Fig. 3.
The part of subsystem, fulfilling the intelligent interface, provides the presentation of user's requirements to the form understandable to integrated part of the system which is actually a knowledge base. The solver is the aggregate of means providing synthesis (automatic in ideal version) of programs of tasks decisions.
The integrated part takes the central position as far as the other components of processing subsystem is concerned and it carries out the interactions of all other parts of subsystem. The conceptual level of integrated part serves to information transformation being in enquiry to the form available to operate by database management system (DBMS) and GIS (informational level). Such dilevel arrangement of inquiry processing of the users provides the effective presentation of knowledge on the top level and the data on the bottom. The subsystem division into three parts (intelligent interface, integrated part, actual system) provides the independence of adaptation means to the final user, adaptation means to application domain and adaptation means to the type of tasks.
Actual part represents the amount of application programs implementing different algorythms of information processing.
Fig. 3 shows the interaction of subsystem parts of processing and presenting in tasks solving of process simulation of pollution spreading in different environments.

3 DATABASES, SOFTWARE, MATHEMATICAL MODELS

To the present there are solving the following tasks:
- data bases are formed about the pollution sources of surface waters and air in large cities of Republic
- data bases are formed about chemical substances available at the enterprises
- data bases are formed about the pollution of the Bashkortostan territory by dioxin and heavy metals
- data bases are formed about huge hydrotechnical installations located on the territory of Republic
- data bases are formed about big emergencies at the industrial enterprises and transport during 1989-1992
- data bases are formed about sick rate and death rate of population during 1973-1992
- it is created a geoinformational model of the Republic territory, regions and large cities (among GIS ARC/INFO)

- there are developed mathematic models and software to simulate the dynamic of process of contamination spreading in surface waters and air according to the data about the volumes of discharges
- there is developed the software for assessment of pollution levels of the territory by different substances in accordance with the data of measurements of contamination levels in separate places
- there is developed the applied software to study the airspace pictures showing the state of contamination in large cities of the Republic
- there are developed the special statistic methods of estimation of possibility of emergency occurrence at technical installations based on prior data such emergencies (methods of small samples processing)
- it is created the software frame for integration into the entire simulation system of previously formed software products destinated for simulating of different ecological processes with showing of simulation results on electronic maps.

Some results of analysis and simulation of ecological situation on the territory of Bashkortostan Republic in EIMS are shown below.

4 BIOGRAPHY

Rustem Z. Khamitov received a Ph. D. in technics from the Ufa Aviation University in 1984. IIis research interests include mathematical modeling of environmental management and planning.

Vladimir E. Gvozdev received a Ph. D. from the Ufa Aviation University in 1983. His research interests include the design and implementation of computer systems to support environmental management and planning, special methods of applied statistics, oriented to small volumes of data.

Sergei V. Pavlov received a Ph. D. in technics from the Ufa Aviation University in 1983. His research interests include the design and implementation of distributed data bases and computer systems to support environmental management and planning.

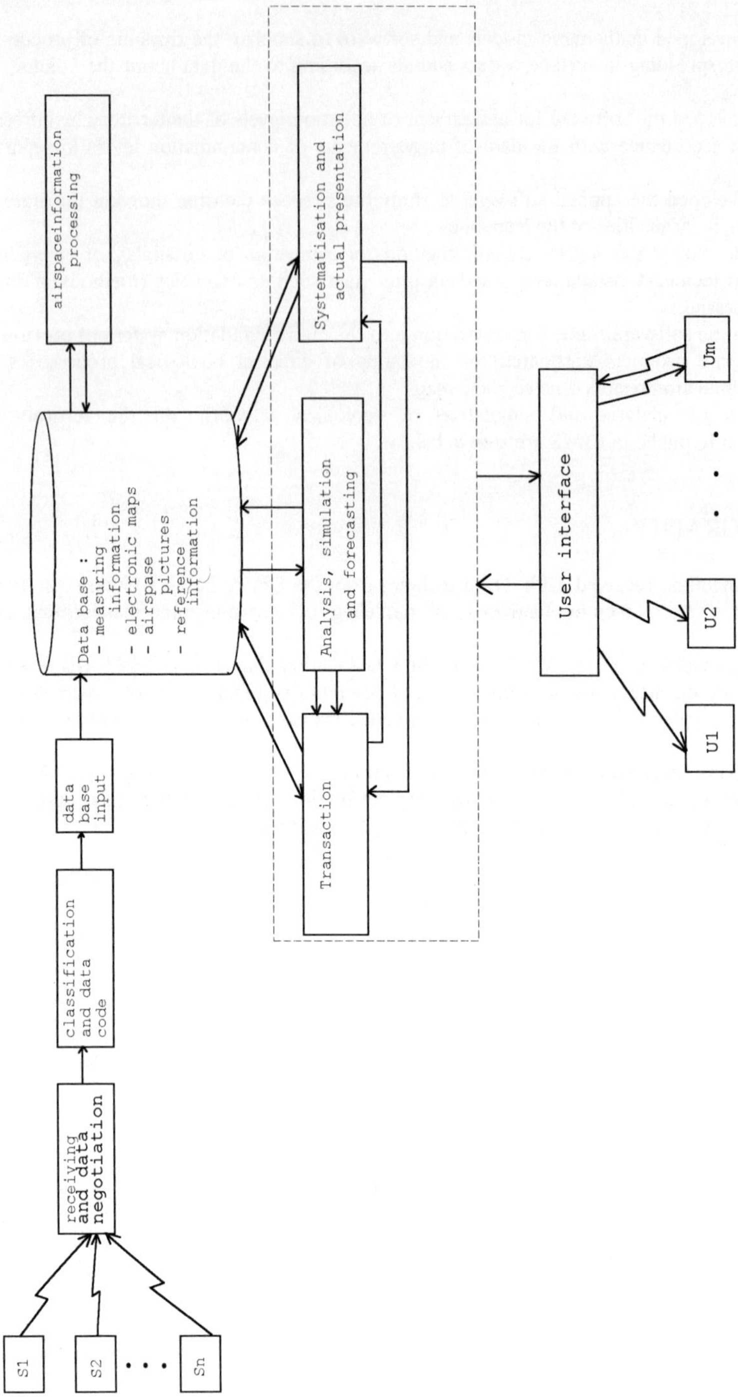

Figure 1 Functional diagram EIMS

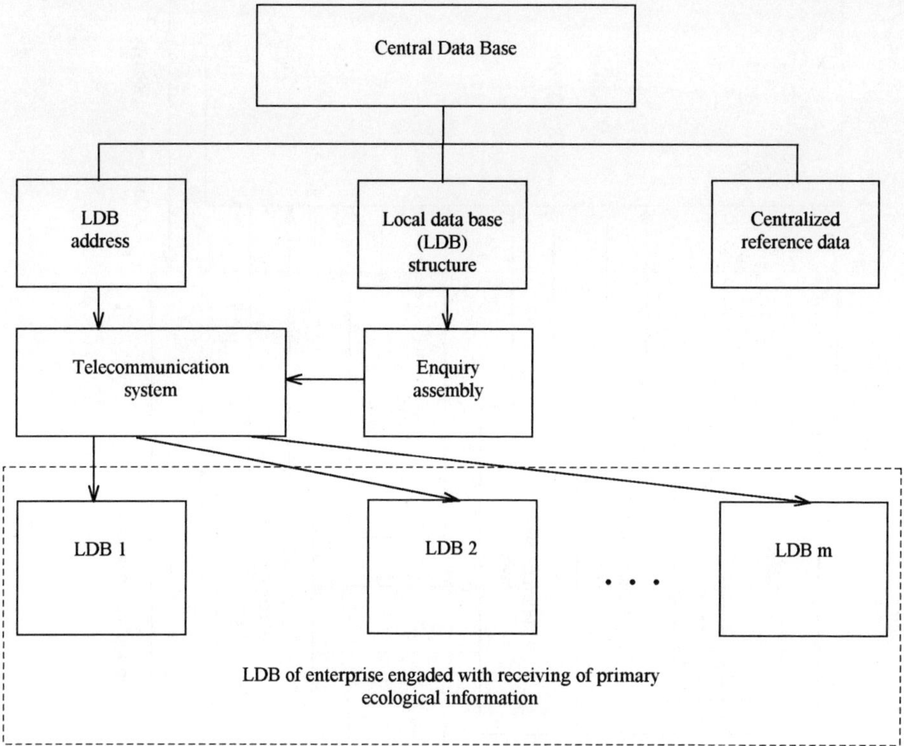

Figure 2 Arrangement of information exchange in the distributed data base

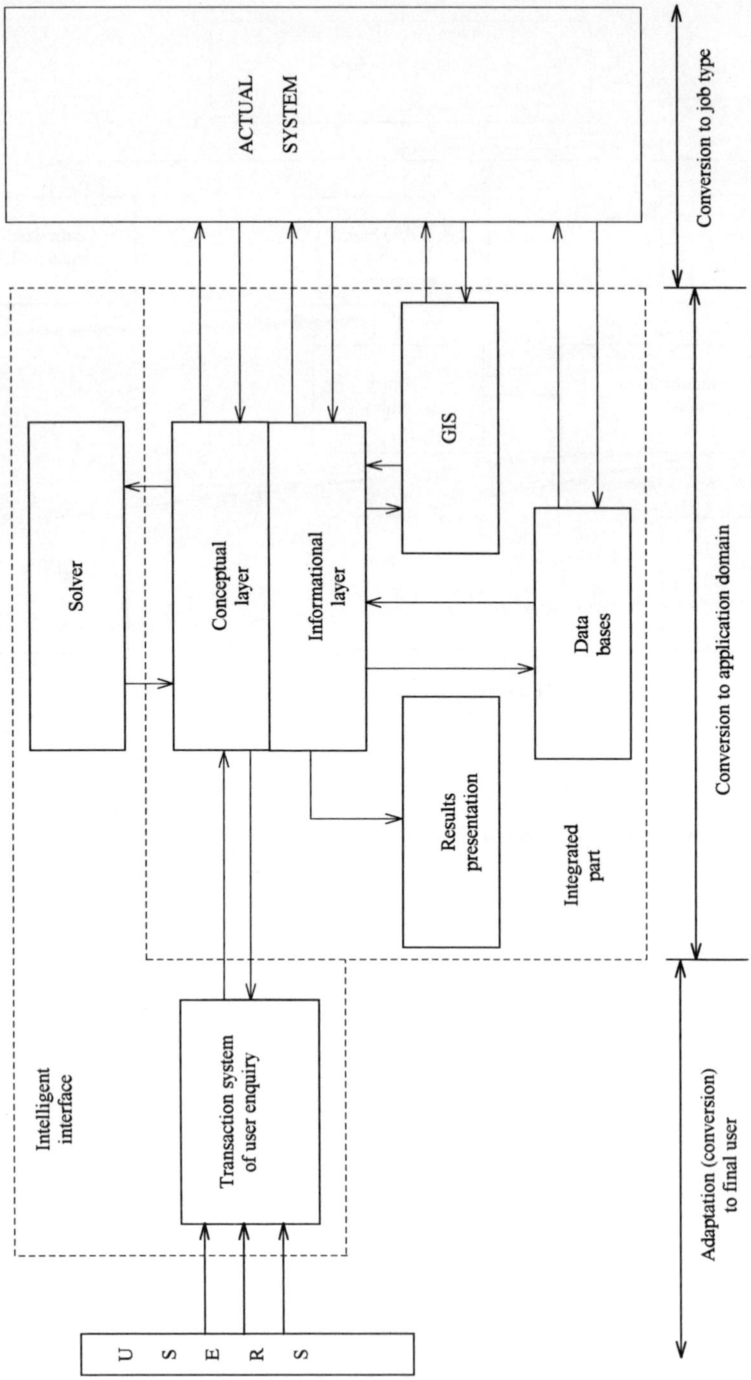

Figure 3 Subsystem stucture of processing and presentation.

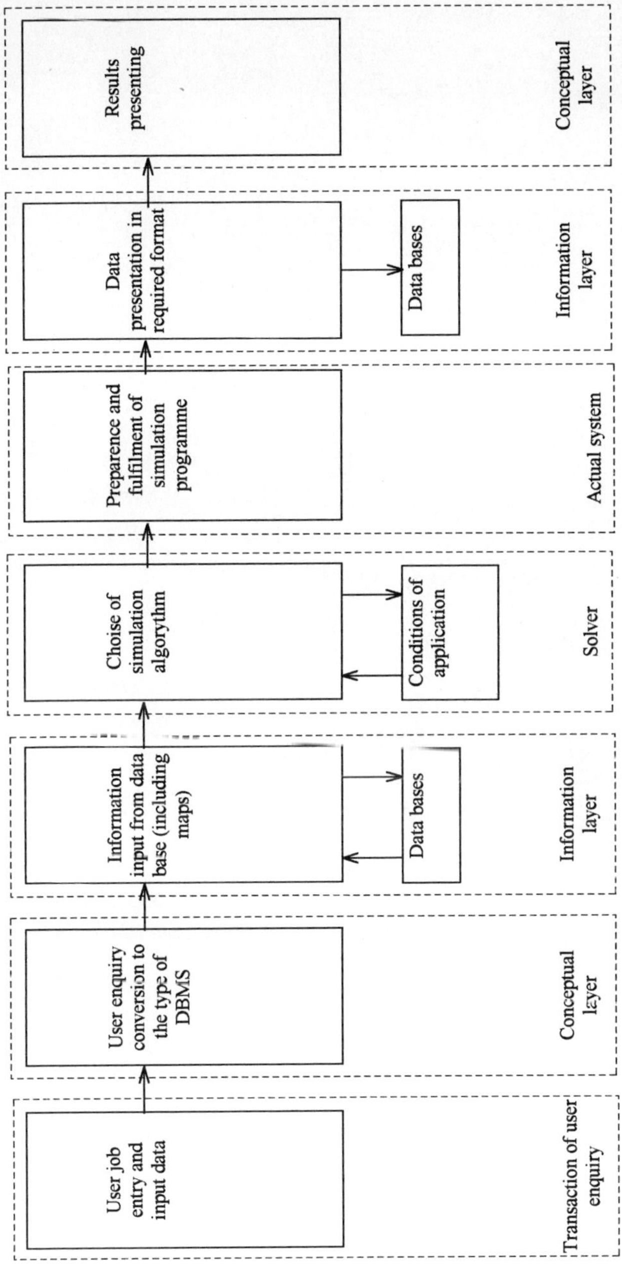

Figure 4 Relationship between subsystem parts of processing and presentation in solving simulation tasks.

PART THREE

Modeling and Simulation

9

Environmental modeling and simulation - applications and future requirements

R. Grützner
University of Rostock
Department of Computer Science,
Albert-Einstein-Str. 21, D-18059 Rostock, Germany
Tel.:+49(381)498 3369, Fax: +49(381)498 3426
e-mail: gruet@informatik.uni-rostock.de

Abstract

This paper gives a short report on the state of the art in the method of problem solving by modeling and simulation used in environmental protection. Application fields, problems, and future developments will be discussed. Requirements and new approaches to the software tools for model and experimental description, simulation system architectures for parallel and distributed simulation experiments will be represented.

Keywords

modeling, simulation experiments, simulation system architecture, environmental applications

1 BASE DEFINITION

Up to now the state of the environment has been changed by human activities in a dramatic and dangerous way. This situation must be stopped as soon as possible. Therefore, we need innovative theoretical and practical scientific approaches. The urgency of the problem poses a challenge to computer science and particularly to environmentally oriented informatics. Important methods and tools are modeling and simulation. Modeling and simulation is a problem solving method where problems are solved not by experimentation with real world systems but by experiments using models of real world objects. Using this method, the main task is to develop adequate models of the real world systems which we want to examine. If a simulation model adequately represents the important aspects of the real system, selected for investigations, a conclusion is then permitted from the simulation results to the behaviour of the real system. Therefore, modeling and simulation is a very important method in practical and theoretical environmental research and protection.

To solve a given problem every model pursues a special aim - the model or simulation aim - which defines the scope of the model. Only these components of a real system are mapped in a model which are significant for the defined model aim. That means modeling is a process of abstraction and generalization neglecting all unimportant real world system components (without any contribution to the model aim).

Simulation models are abstract mathematical descriptions of system relations, the relations between the states of a system. For descriptions model definition languages (e.g. DYMOLA) or simulation languages (e.g. SIMULA, ACSL) are used.

Depending on the value scope of the state variables given, we distinguish models or systems as continuous, discrete or combined. Models are called continuous if the variables of the system possess values from an intervall, discrete if system variables assume discrete values. In combined models system variables assume discrete and continuous values.

To get simulation results, we have to experiment with models. In an experiment computation methods are applied to models (models are executed by a method). Computation methods or short methods are computation algorithms. A base method is the simulation method which controls the time-dependent dynamical behaviour of the system state. Other examples of methods at a higher level are optimization of parameters or system trajectories, sensitivity analysis, and Monte-Carlo methods (stochastic changing of parameters or inputs).

Experiments require an appropriate experimental description to define the kind of problem solving needed. If model and experimental descriptions are separated, we get an important base for a multiple use of models.

Essential software systems for modeling and simulation activities are simulation systems. A simulation system supports the user during the phases of model building and description, the experiment definition process, the execution and supervision of simulation experiments, the analysis and visualization of simulation results, the managment of models, data, and experimentation methods.

2 MODELLING AND SIMULATION OF ENVIRONMENTAL SYSTEMS

Modeling and simulation in the environmental domain today is used to determine the dynamic behaviour of systems in order to make decisions or to test the impact of human actitivities (e.g. the use of solar energy for heating, construction, production processes, traffic and logistic planning and decisions) on the environment. Further questions will be added in the near future because of the increasing complexity of the systems: the assessment of alternative decisions and activities, the examination of sceneries, the analysis of eco-economical and of socio-economical systems. Additional future assignments would be the evaluation and optimization of the whole life-time cycle of products related to their impacts on the environment. The life-time cycle of a product consists of its production, use, and finally, the recycling process.

To solve these assignments, new experimentation approaches, simulation methods, computation methods, and software tools are needed. Several of the application fields and some software architectures for modeling and simulation will be represented in the following.

2.1 Problems of Modeling Environmental Systems

The modeling and simulation of environmental systems requires the consideration of a lot of special system properties which complicate the analysis considerably.

We have to consider different groups of problems. One group results from the complexity of the environmental systems (e.g. the biological, ecological, economical properties) and the other group results from the degree of the user's qualification in managing these problems (using the models, storing and preparation of data, interpretation of results and, the making correct decisions).

Problems of modeling environmental systems:

- there is a lack of theoretical knowledge
- causal relations between system components are frequently unknown
- most of the real systems can be observed by experiments, but frequently there isn't any possibility for controlled experiments
- ecological systems very often possess long reaction times (up to years or centuries) and in contrast, other components possess very short reaction times
- system complexity is high because of the strong relations between the system components
- nature is an open self-organizing system with adaptable components
- inaccurate data and ill-defined system components exist
- there is an increasing set of data: spatial data and spatial data with time dependence.
- often it isn't possible to define exact and unique optimization aims for ecological and environmental systems
- the model description must be realized by analytical and non analytical (rule based) methods.

Another group of problems result from the different levels of knowledge by users in the field of environmental protection. There are highly qualified specialists without any knowledge in the field of simulation and in the use of complex software systems. This situation requires software with excellent user interfaces and user support components for the whole model-life-time cycle. Unfortunately, we can't very often find simulation software which possess only some of these expected support components. Such systems are still in research.

2.2 Fields of Application

The most important tasks in the field of environmental modeling and simulation today are still the determination and analysis of the behaviour of environmental systems, that means the computation of the time dependent dynamic behaviour of state variables.

Some applications of modeling and simulation are given below. On the one hand, undesired secondary effects of human activities (e.g. emissions of harmful substances of production processes) can appear. In such a situation, the secondary effects to the environment are to minimize. On the other hand, actions of environmental protection are to optimize. These are investigations of the questions: how does a process work? or which effects are released in the environmental system by a special input?

In the environmental domain, simulation models are especially used in four fields: emission computation (air, water, ground pollution), process control, groundwater - economical and flow investigations, and ecosystem research.

But further applications become more and more important: e.g., models on the use and balance of resources (e.g. water, ground, materials, energy, food); models of the carrying capacitiy and of loading limits of ecological systems with an input, which has been caused by human activities; quality models; product-life-time-cycle models and ecobalances; socio-economic models; combined tasks: the use of simulation and optimization methods or experiments consisting of coupled economical and ecological models.

Models of pollution extension

An important task in environmental protection is the computation of immission loads caused by emission processes. Most environmental problems are caused by the emission of harmful and dangerous substances which are spreaded in air, water, and on the ground. The computation of the expansion of radiation, noise, vibrations, chemical reactions of substances and their effects on human life are of high interest today.
The primary targets of extension models are:

- computation of pollution values at geometrical points where we don't have any measurement results
- detection of pollution sources by simulation experiments
- supporting position planning and permission processes for objects with emissions (e.g. of factories, production processes, traffic projects, buildings)
- determination of reactions to accidents (e.g. of chemical processes, of oil transports).

An overview of the state of the art is given by (Zannetti,1990), (Wrobel,1991), (Melli,1992). Models are urgently needed in the future for the computation of pollution and distribution in the micro scale range (near the ground). For example, there are problems regarding the distribution of auto exhaust in street canyons depending on traffic density and kind of formation of buildings near the street.

Water models

The pollution and the increased use of water based on constant water resources has produced evident environmental damage all over the world. This causes a high importance of water models. There are three main groups of models:

- investigations of the influence of structures, buildings and other human activities on the amount and flow of groundwater
- investigations of the levels of groundwater depending on irrigation, draining, and soil erosion.
- waste water models and water cleaning processes.

An important future assignment will be the modeling and simulation of the relations between vegetation and groundwater levels .

Models for process control simulation

Models are used to evaluate the consumption of recources of technical processes and the emission of these processes. The primary target is the minimizing of the environmental influence. Models are used during the planning phase and the model supported process control to get an optimal process behaviour. Examples of process models are given by (Keller, 1994), (Tuma, 1994), (Früh, 1989).

Modeling of ecosystems

The two fields of environmental protection and ecosystems are strongly connected. But today a broad use of simulation is still difficult. Simulation is used only in selected parts, e.g. community ecology in water, ground, forest; nitrogen balance; steadiness and elasticity of ecosystems - that means the capability of taking up harmful substances without any effects on the ecological balance; effects of chemical substances in ecological systems.

More than in other applications, modeling and simulation is an interdisciplinary task. The following part of this paper will refer to the modeling and simulation of ecological systems connected with environmental protection.

2.3 Problems and Expected Results

Modeling and simulation cause some problems when using existing simulation systems and simulation software tools. The main problems are:

- it is necessary to get a better understanding of the dynamic behaviour of the systems by using structure edaquate models
- missing system support for investigations of human processes in such a way, that damages to the environment will be at a minimum
- handling the increasing system complexity (based on extensions of system borders and/or detailed system investigations) by modular-hierarchical modeling concepts
- development of comfortable user interfaces of software tools for simulation experiments.

An improvement in the present situation can be achieved on the one hand by the building of more precise models (e.g. of structure adequate models, (Bossel,1994)) and on the other hand by the development of better simulation software. We will explain the last point below.

The complexity of environmental systems demands a model description so that the method of description can be adapted to knowledge about the system. Models consisting of different described modules may be called *heterogenous*. For instance one modul of the model may be described by differential equations and other moduls by rules, difference equations, higher Petri-nets (ecocontroling models), or other tools. The simulation systems must be able to simulate such heterogenous models. That means, that dependent on the description method of a module appropriate simulation methods must be used. The structure of the simulation system must be changed to realize simulation runs of heterogenous models.

An effective experimentation requires control of experiments either interactive or by an experiment description. Experimental control means to fix the methods, sequence of use, and data to solve a problem in a simulation experiment. Methods are used for optimization, sensitivity analysis, investigation of transient states, steady-state detection, visualization, and so on.

3 MODERN STRUCTURES OF SIMULATION SYSTEMS

The following considerations are restricted to the modeling and simulation of ecological problems, that means the consideration of combined systems which are described by systems of ordinary differential equations of first order (system state equations).

Computer science can support the modeling and simulation in three important tasks:

- building and managment of models and experiments
- supervision of simulation experiments by adequate simulation system architectures
- simulation result analysis and visualization.

The first and second points are investigated in our projects SAME and SAMEC (advanced by the Deutsche Forschungsgemeinschaft), (Grützner,1993).

3.1 Models and Modeling

One possibilty in handling the system complexity and to guarantee the multiple use of models is the separation of model and experiment description (Schwarze, 1990), (Zeigler, 1986), (Zeigler,1989). Up to now this has been realized in only a few systems.
Separation means that a model represents only the structural description of a class of systems. All of the values of parameters, of initial state variables, and system input variables are defined in the experimental description belonging to the model. Aside from the values an experimental description also defines the operations (or methods) which are applied to the model (e.g. optimization of parameters) during the experiment's execution. The experiments control the model and the method execution during the simulation process.

But there is a disadvantage in the separation because we need two languages, which a user must know; one language for the model and another one for the experimental description.

Present model description languages are simulation languages belonging to a special simulation system. This is a disadvantage because we are unable to exchange models between different simulation systems and the user has to learn another simulation language if he wants to use a new simulation tool. For the considered field of application, it seems important for the future to look for a general modeling language independent of any simulation system. Such a general language will allow a better model exchange and system understanding. To do a simulation currently, the model description would have to transform into the language used by the system. Such an approach has been used successfully by electrical and electronical model descriptions. The proposed language VHDL-A will be standardized soon (Bausch-Gall, 1994).

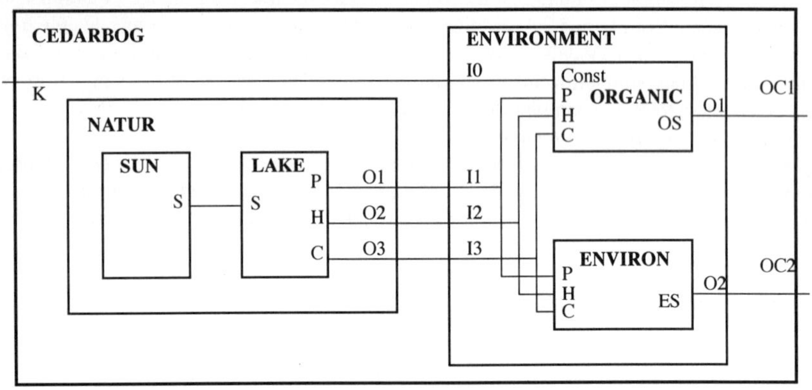

Figure 1 Graphical description of the modular-hierarchical model CEDARBOG.

Modell Description

The models in the environmental domain are modular-hierarchical (a second concept to handle the system complexity) and of a variable structure. The systems are decomposed into *base objects.* To find a solution to an existing problem, a further decomposition of a base object isn't necessary. Base objects contain the equations for the dynamic behaviour of the system. Coupled base objects are called *submodels.* These can be coupled again to get submodels at a higher level. So a model can be used as a submodel in another model. This means object oriented modeling, where a base object represents a real system object.

It is important for users to have a model description language and a graphical modeling tool at the same time. A mix of both descriptions in one model should be allowed.

The graphical model description of a simple example CEDARBOG of a silting-up process in a lake see Figur 1 and the description by the language SAME-MDL see Figur 2. The model consists of two submodels, NATURE and ENVIRONMENT, which consist of two base objects. SUN defines the energy input by the sun, base object LAKE defines the state of plants (P), of herbivorous objects (H), and of carnivorous objects (C). The base object ORGANIC

Basicobjects

```
basicobject SUN(out:S);
  constants float PI=3.14;
  var double S=0.0;
  equations
    S = 95.9 * (1 + 0.635*sin(2*PI*TIME));
end.


basicobject LAKE(inp:S; out:P,H,C);
  states float P=0.0, /*Plants*/
               H=0.0, /*Herbivorous objects*/
               C=0.0; /*Carnivorous objects*/
  var double S=95.9;
  equations
    P'= S 4.03 * P;
    H'= 0.48 * P -17.87 * H;
    C'= 4.85 * H - 4.65 * C;
end.


basicobject ORGANIC(out:OS; inp:P,H,C,Const);
  states float O=0.0;
  var float P,H,C;
  equations
    OS = O * 0.1;
    O'= Const * P + 6.12*H + 1.95*C;
end.


basicobject ENVIRON(out:ES; inp: P,H,C);
  states float E=0.0;
  var float P,H,C;
  equations
    ES = E * 100;
    E'= 1.00 * P + 6.90*H + 2.70*C;
end.
```

Submodels

```
submodel NATUR(out: O1,O2,O3);
  components
    SUN(out: S);
    LAKE(inp: S; out: P,H,C);
  connections
    SUN.S -> LAKE.S;
    LAKE.P -> O1;
    LAKE.H -> O2;
    LAKE.C -> O3;
end.
submodel ENVIRONMENT(out: O1,O2;
                     inp: I0,I1,I2,I3);
  components
    ORGANIC(out: OS;inp: P,H,C,Const);
    ENVIRON(out: ES; inp: P,H,C);
  connections
    I0 -> ORGANIC.Const;
    I1 -> (ORGANIC.P,ENVIRON.P);
    I2 -> (ORGANIC.H,ENVIRON.H);
    I3 -> (ORGANIC.C,ENVIRON.C):
    ORGANIC.O -> O1;
    ENVIRON.E -> O2;
end.
submodel CEDARBOG(inp:K; out:OC1,OC2);
  components
    NATUR(out: O1,O2,O3);
    ENVIRONMENT(out:O1,O2;
                     inp: I0,I1,I2,I3);
  connections
    K -> ENVIRONMENT.I0;
    NATUR.O1 -> ENVIRONMENT.I1;
    NATUR.O2 -> ENVIRONMENT.I2;
    NATUR.O3 -> ENVIRONMENT.I3;
    ENVIRONMENT.O1 -> OC1;
    ENVIRONMENT.O2 -> OC2;
end.
```

Figure 2 Model description of CEDARBOG using SAME-MDL, (Dimitrov,1994).

models the sediment process and ENVIRON the energy output to the environment.

Because of the above mentioned problems in environmental modeling, it's important to allow the use of different mathematical descriptions of the dynamic behaviour of the base objects, e.g. by differential and difference equations, rule based approaches, fuzzy rules, and neuronal net model components. Up to now this hasn't been realized by SAMEC.

Experimental Description

The proposed procedural experiment description language (Dimitrov,1994) is based on the frame concept. A frame is a structuring element for an experiment also to define subexperiments and a parallel experiment execution. Subexperiments can be described depending on additional conditions. One example of such an experiment description is given by Figure 3. This is an experiment with the model CEDARBOG for two parallel experiments using this modell. CEDARBOG may be used with two different parameter sets.

```
experiment EXAMPLE;
variables  float konst;
           timeseries OC11, OC21, OC12, OC22; /* result files */
frame
        frame1 (model_name, inparam, result1, result2, tstart, tend);
        {simulate(model_name; RK4,tstart,tend; inparam,result1,result2)};
        frame2 (model, inconst, res1, res2, tstart, tend);
        {konst = inconst;
        simulate (model; RK4,tstart,tend; inconst,res1,res2)};
control
        parbegin
          frame1 (CEDARBOG, 2.55, OC11, OC21, 0, 250);
          frame2 (CEDARBOG, 5.10, OC12, OC22, 0, 250);
        parend
end.
```

Figure 3 Description of two parallel experiments with the model CEDARBOG.

3.2 Simulation Systems

Let us consider simulation systems for ecological systems. The architecture of simulation systems is affected by the following requests:

- interactive use in all working phases and user friendly interfaces
- support of the users in the building and managment processes of models and experiments
- parallel and distributed execution of experiments
- use of innovative methods: genetical algorithms, fuzzy set and fuzzy rule based concepts, neuronal nets
- coupling of simulation systems with information systems (geo-information systems).

Parallel simulation

Based on the modular-hierarchical model concept, we have realized a system architecture for parallel and distributed simulations of base objects. The system of algebraic-differential equations is separated into subsystems, the equations of the base objects. For the parallel simulation of the base objects special integration algorithms have been used (Schäfer,1994), (Werner,1995). Another iterative algorithm for parallel simulation, which uses the classical numerical integration methods for base objects, is proposed by (Singh,1986). But in contrast to the first algorithms the iterative solution is inaccurate.

To control the couple relation between the base objects and the submodels special coordinator routines are introduced. A coordinator realizes the data exchange between the model components and some functions for the integration process. The external structure of the model is in this architecture equivalent to the internal structure. Some results of measurements neglecting the communication overhead indicate an evident speed up (Werner,1995).

Parallel experiments

An experiment description also allows a parallel experiment realization. Each experiment is then controlled by its own experiment process. Parallel experiments are especially important (high speed up) for parameter and trajectory optimization, sensitiveness investigations, and other multiple simulation runs.

The communication between processes are based on the PVM concept (Parallel Virtual Machine). A special communication **p**rotocol of **s**imulation system **p**rocesses (PSP) has been defined for implementation (Nekien,1995). The implementation has been realized at an UNIX workstation net.

3.3 Coupling of Simulation Systems with other Software Tools

Modeling and Simulation of environmental systems means working with a lot of data (e.g. input, geographical, environmental, result data). For an effective simulation all data must be provided automatically from different data bases (e.g. geographical information systems, environmental information systems). The access to the data supposes a coupling of simulation systems and data bases. The data which are used in the simulation experiments must be defined in the experiment description in such a way, that no manual actions by users are necessary. The access description to data bases in experiments must be possible without any knowledge of internal data base structures.

4 REFERENCES

Bausch-Gall,I. (1994) Modellaustausch zwischen Simulationsprogrammen. in *Fortschritte der Simulationstechnik, Bd. 9, ASIM 9. Symposium* (ed. G. Kampe & M. Zeitz), Vieweg, Stuttgart.

Bossel, H. (1994) Understanding Dynamic Systems: Shifting the Focus from Data to Structure, in *8. Symposium Informatik für den Umweltschutz* (ed. L. M. Hilty, A. Jaeschke, B. Page, A. Schwabl .), Bd 1, Metropolis Verlag, Marburg, 63-75.

Dimitrov, E. (1994) *Modell- und Experimentbeschreibungen für Umweltproblemstellungen und ihre Implementation* . Preprint No. CS-01-94, University of Rostock, Dept. of Computer Science, WG Modeling/Simulation.

Früh, K. F. (1989) Informationsorientierte Prozeßleittechnik für die Chemische Industrie. *Automatisierungstechnische Praxis atp*, **31**(7), 301-305.

Grützner, R. (1993) Abschlußbericht zum Projekt SAME. University of Rostock, Dept. of Computer Science, Report of the WG Modeling/Simulation

Keller, H. B., Kugele, E., große Osterhues, B. (1994) Einsatz neuronaler Netze im Umweltbereich am Beispiel der Müllverbrennung. in *8. Symposium Informatik für den Umweltschutz* (ed. L. M. Hilty, A. Jaeschke, B. Page, A. Schwabl .), Bd.1, Metropolis Verlag, Marburg, 405-414.

Melli, P., Zannetti, P. (1992) *Environmental Modelling*. Computational Mechanics Publications Southampton.

Nekien, T. (1995) *The communication protocol in the SAME-simulator*. University of Rostock, Dept. of Computer Science, WG Modeling/Simulation, in publishing.

Schäfer, P. (1994) Parallele Verfahren zur Simulation modularer technischer Systeme. in *ASIM 9. Symposium* (ed. G. Kampe & M. Zeitz), Vieweg, Stuttgart.

Schwarze, G. (1990) *Digitale Simulation*. Akademie Verlag, Berlin.

Singh, M. G., Allidina, A. Y., Daniels, B. K. (1986) *Parallel Processing Techniques for Simulation*. Plenum press, New York, London.

Tuma, A., Haasis, H. D. (1994) Real Time Production Scheduling of Environmental Integrated Production Systems - A Comparision of Selected Knowledge-Based Methods and Maschine Learning Algorithms. in *8. Symposium Informatik für den Umweltschutz* (ed. L. M. Hilty, A. Jaeschke, B. Page, A. Schwabl), Bd.1, Metropolis Verlag,Marburg, 371-8

Wrobel, L. C., Brebbia, C. A. (1991) *Water Pollution: Modelling; Measuring and Prediction*. Computational Mechanics Publication, Southampton.

Werner, G.(1995) Vergleichende Analyse von parallelen Simulationsmethoden zur Ausführung komplexer Modelle. University of Rostock, Dept. of Computer Science, Diplomarbeit.

Zannetti, P. (1990) *Air Pollution Modelling*. Computational Mechanics Publication, Southampton.

Zeigler, B. P., Oren, T. I. (1986) Multifacetted Multiparadigm Modelling Perspectives: Tools for the 90's. in *Winter Simulation Conference*. SCS Publications, Washington , 708-712.

Zeigler, B. P., Zhang, G. (1989) The Entity Structure: Knowledge Representation for Modeling Design. in *Artificial Intelligence, Simulation, and Modeling* (ed. L. E. Widman, K. A. Loparo, N. R. Nielsen), John Wiley&Sons, New York, 47 - 73.

5 BIOGRAPHY

After a study of applied aerodynamics and an activitiy of seven years in the industry follows an activity as research worker at the Humboldt University of Berlin, fields of modeling and simulation, computer performance analysis, and operating systems. Since 1989 professor for modeling/simulation at the University of Rostock. Working fields: architecture of simulation systems, parallel simulation, user interfaces, application of simulation in environmental research.

10

A Framework for Distributed Processing of Air Pollution Dispersion Models

Gerhard Eschelbeck, Christian Hartl
Research-Institute for Microprocessor Applications (FIM)
University Linz, A-4040 Linz, Austria
Email: eschelbeck@fim.uni-linz.ac.at

Abstract

The increasing pollution of our atmosphere due to organic and inorganic emissions from industry and a lot of factories require the availability of reliable but simple tools for the calculation of air pollutant emissions. Many different calculation and simulation models are approved nowadays. One difficulty is the increasing requirement of computational power. Our idea was to implement a general architecture for parallel graphical pollution simulation providing the integration of new calculation models in a very fast way. We chose an object oriented approach, which allows to integrate further calculation models and special user interface components in an efficient way. Thereby, the user can simulate environmental situations due to various air pollution dispersion models without changing the application, but by selecting the corresponding calculation model. The benefit of our solution is the separation of the input unit from the calculation and from the output unit. For reducing the system requirements on Windows / Windows NT environments, the calculation unit is based on a distributed client / server architecture. It deals with dynamic job distribution within networks for parallel computation without the need of special workstations. Thereby free resources are used for increasing performance. Standardized remote procedure calls are the basic mechanism for distributing jobs to remote computers. For analysing and benchmarking our framework we have implemented a calculation module for the approved standard OENORM M9440.

Keywords

Pollutant, Dispersion, Parallel Simulation Model, Client/Server, Distributed Calculation

1 INTRODUCTION

It is not our competence to discuss the amount of dispersion which is harmful or not for the mankind, that's the medical and toxicological experts' job. We will show how to process data and information of concentrations and distributions, which support government for their

decisions. Due to low resource requirements our application opens for a large group of potential users. It gives the possibility to simulate and visualize various situations for future industry settlement planning and licensing of emittants. Furthermore it can be shown how improvements like integration of filters reflect on the emission volume.
Our application consists of an essential main component and a lot of different optional calculation modules which can be linked to the base calculation unit as add-on parts. The overall system bases on an object oriented architecture fulfilling modelling and simulation of different environmental locations under varying situations. To improve performance for simulation the base calculation unit is responsible for job distribution to remote computers depending on available resources. Partitioning of the model into subjobs must be done by the corresponding calculation module itself. The application allows to define situations with individual sources and receptor points modelling the particular environment. Receptor points are the geographical points indicated by the spatial coordinates x,y and z, for which the concentration of air pollutant should be determined. Sources are places where the exhausted gas - a mix of various substances - is emitted from plants into the atmosphere. We have integrated an interface for loading predefined information about local environment and geographical data (cities, mountains, streams, ...) from external files. This allows to setup background information for better preparation of outputs.

2 A GENERAL FRAMEWORK FOR PARALLEL SIMULATION AND VISUALISATION OF ENVIRONMENTAL INFORMATION

As mentioned above and given in figure 1 our architecture is divided into mainly three components (input, calculation and output unit). The calculation unit is subdivided into the mandatory base unit and optional calculation modules specifying various simulation models.

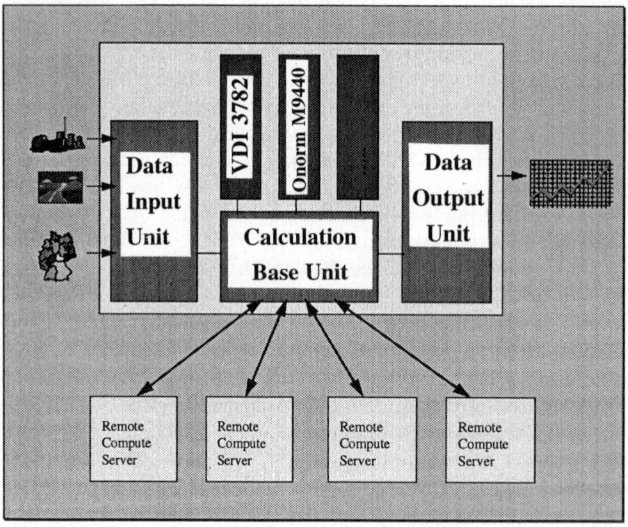

Figure 1 Architecture of our solution for simulation and visualisation of environmental information with job distribution.

Graphical user interface and input of common data is done by the input unit. Preparation of simulated data for graphical and tabular output is task of the output unit. Due to our modular architecture the specific simulation models are optional, depending on individual needs. These optional units are built by inherited classes which usually exist of additional user interface components as well as the algorithms for simulation. Further input information needed for special air pollution dispersion models are handled by itself. We organized them as dynamic link libraries, a sample for a special approved austrian standard is given in chapter 3. The modular structure guarantees flexibility for adding different kinds of air pollution dispersion model (i.e. OENORM 9440, VDI 3782, Dutch model,..) and increases the reuse of the complex and powerful architecture. With respect to the compatibility of the different dispersion models we defined a generic interface between calculation base unit and the optional calculation modules. The modules are represented as object which are derived from the virtual base class and embedded into a dynamic link library. All the specific dialog boxes and necessary input data are handled by the themself. Therefore this architecture allows to adopt future standards very easily.

2.1 Input Unit

We handle different types of input information which can be entered by the user in different ways. These various forms of information build the basis for further calculation steps. The following list gives an overview of the basic input parameters required to characterize individual emission situations:

Information about sources

- position and dimensions (height, diameter) of the source
- list of emitted substances (i.e. SO2, CO, NO, NO2, H2S,..)
- mass flow of the emitted substances
- exhaust volume rate
- exhaust temperature at source height

Information about receptor points

- coordinates of the receptor point

All the emission sources defined by the user for a special situation contain flags for individual controlling activity during simulation, therefore various conditions and views can be evaluated. For configuring this variety of information we support different mechanisms:

- **Geographical background information:** Input of information from external data files is supported by special interfaces for importing different fileformats (BMP, DXF, Bitmaps,..). This background information enables an improved graphical representation and allows easier navigation for the user.
- **Definition of sources and receptor points:** Predefined symbols enable the graphical definition of an environmental situation in the Cartesian as well as the Gauß-Krüger coordinate system. The Cartesian coordinates are relative to a reference point, which is often simultaneously the zero point. The Gauß-Krüger coordinates relate every point of the earth to a definite point with respect to a reference meridian. If the Gauß-Krüger coordinates of all sources and receptor points are known, they could be entered simply.

Source points must be configured additionally with lists of emitted substances and corresponding parameters.

- **Individual drawings:** A set of simple tools (line, box, circle,...) allows to draw individual information or critical areas with a graphical method.
- **Standard parameters:** Special dialogs allow to define various standard parameters for substances and air pollutants.

These tools allow the definition of individual emission environments. Therefore they establish the base for further simulation and the situation can be stored on external files for later reuse or new calculation.

2.2 Calculation Unit

The calculation unit receives it's information from the input unit and forwards results to the output unit for presentation. As the base object of our framework the mandatory base calculation module defines the interface for linking specific calculation modules. Additional models represented as dynamic link libraries can be bound. An example of an optional calculation model is given in chapter three. Furthermore the base module establishes the connection to the remote compute servers (see figure 1) which are responsible for job computation. Depending on available resources jobs are distributed to connected computers which are configured for remote processing. Within the calculation unit the specific models are responsible for partitioning and forwarding their jobs to the base unit which distributes them for calculation. Due to parallel calculation we get a speedup depending on the calculation model and on available resources on remote servers[Kum94] also. Figure 2 gives the benchmark results for the OENORM M9440 model which leads to a nearby linear speedup. The values for speedup depend on complexity and properties of the model as well as the available resources within the network.

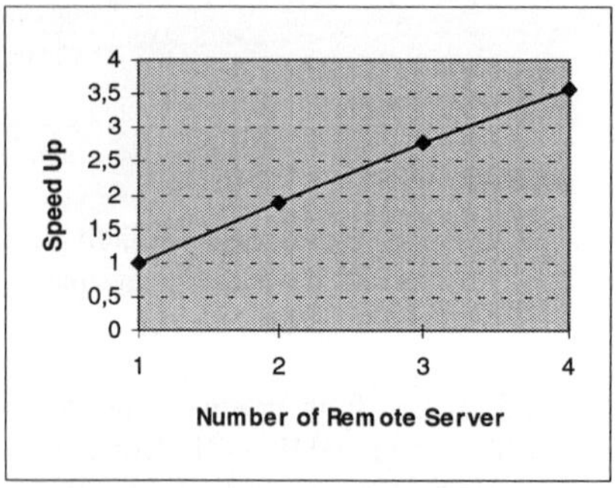

Figure 2 Speedup for parallel processing of the OENORM M9440 model.

The client / server communication between the base calculation unit and the connected remote servers is done by remote procedure calls[Cor91]. We have implemented mechanisms on the remote servers allowing performance and load evaluation. Compute jobs can be distributed to those machines, having free resources. Also the computer running the main application can take part in servicing calculation jobs. Therefore also local systems can execute our application and no minimum number of remote machines is required. The maximum number of computers for parallel calculation is limited by the increasing network traffic and the structure of the calculation model also. Time consuming jobs with a small amount of communication parameters are an indicator for gaining high speedups due to parallel processing with our architecture.

2.3 Output Unit

The output unit is a mandatory component which handles graphical and tabular representations as well as density charts. Different views to the results depending on configuration are available:

- Considering sources as stand alone - for viewing and comparing the effects with other sources
- Including all active sources within the emission area and calculating corresponding output

3 CASE STUDY: INTEGRATION OF THE OENORM M9440 STANDARD

Nowadays a wide range of calculation and simulation methods for dispersion models are known and standardized. Examples are the Austrian OENORM M9440, the German VDI 3782 or the Dutch model. We decided to implement the Austrian OENORM M9440 [OEN92] as the first air pollution dispersion model for our simulation framework. This standard is based on the Gaussian dispersion model[OEN92],[VDI92],[VDI91], which simulates the process of dilution and transport of emissions. Therefore, when applying this model, the emission data as well as the meteorological parameters must be known. The advantages of the Gaussian dispersion model compared to more complex models consist in the fact that the dispersion parameters used have been derived from dispersion investigation experiments. The Gaussian model contains approximations and simplifications which sometimes lead to systematic errors. In addition the inaccurate measurement of all input parameters can lead to errors in the calculated concentrations also. For these reasons, the Gaussian dispersion equation represents a dispersion model which has been validated empirically by dispersion experiments. Compared to more complex models, a relatively short computing time is needed.

Our implemented calculation module requests additional input from the user covering some basic meteorological parameters:

- Transport velocity of the wind
- Transport direction of the wind
- Dispersion categories
- Dispersion parameters
- Mixing layer height

One assumption for the correctness of this model is that these parameters are constant over the global emission for a specified period. Only this assumption guarantees the ability to use this calculation model.

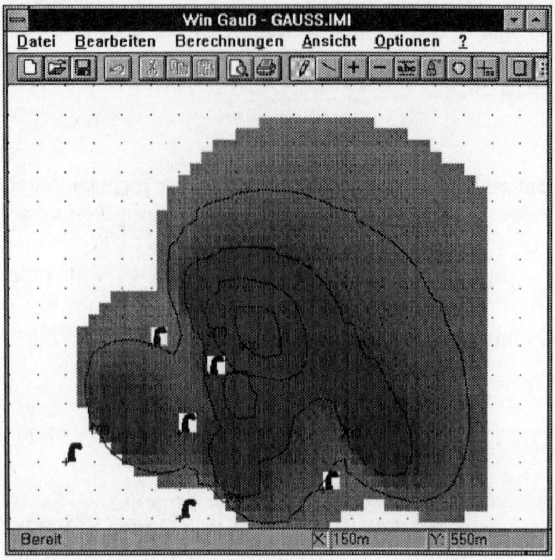

Figure 3 The graphical result for simulation of six sources of the air pollutant SO_2 (transport direction 235 degree, air temperature 20°C).

4 CONCLUSION

This article presents the architectural overview of our generic framework for simulation and visualisation of emission dispersion. Special simulation models can be linked into this framework. Therefore easy adaptation and extension is given by wrapping into dynamic link libraries. By partitioning the simulation models we can distribute jobs to connected computers for parallel computation and increasing performance.
Due to low resource requirement, our solution is suitable for desktop calculation in local and network environments. Fast and easy analyzing of emission situations with graphical data representation is the users advantage.

5 REFERENCES

[Cor91] Corbin, John R. The Art of distributed applications. Programming techniques for remote procedure calls. Springer Verlag, 1991

[Esc94] Eschelbeck G., Mühlbacher J.: A Framework of Classes for Distributed Controlling. Proc. EUROMICRO Conference 1994, pp. 240-246 IEEE Press, California, 1994

[Kum94] Kumar V., Grama A., Gupta A., Karypis G. Introduction to Parallel Computing. The Benjamin/Cummings Publishing Company. California, 1994

[OEN92] OENORM M9440. Ausbreitung von luftverunreinigenden Stoffen in der Atmosphäre. Berechnung von Immissionskonzentrationen und Ermittlung von Schornsteinhöhen.

[VDI92] VDI 3782 Part 1. Gaussian Dispersion Model for Air Quality Managment. Dispersion of Pollutants in the Atmosphere. October 1992

[VDI91] VDI 3782 Blatt 4. Ausbreitung von Geruchsstoffen in der Atmosphäre. Umweltmeteorologie. Mai 1991

6 BIOGRAPHY

Gerhard Eschelbeck is a graduate computer science engineer. In 1993 he joined the Research-Institute for Microprocessor-Applications at the University of Linz, Austria. His current research interests include networking and distributed computing. He is a member of the IEEE, the Computer Society, and the ACM and is an active participant in international computer science conferences.

Christian Hartl studied computer science at the Johannes Kepler University Linz, Austria. Main emphasis in the field of system programming, algorithm and data structures. Establishment of the diploma thesis at the Research Institute of Microprocessor Applications with the topic "Dynamic load distribution". Further interests in the field of Computer supported cooperative work (CSCW).

11

Trinity: model-based support for multi-actor environmental problem solving

Henk B. Diepenmaat
TNO Institute of Environmental Sciences
Schoemakerstraat 97, P.O. Box 6011, 2600 JA Delft,
The Netherlands, Phone +31 15 696133, Fax +31 15 616812

Abstract

An environmental problem solver is generally confronted with a complex situation of interacting, and in many cases even conflicting, (intentional) activities. Methodological support for problem solving processes in such problem contexts is lacking. In this paper, a short description is provided of a tool that aims to fill in this methodological gap: the ***Trinity*** approach.

Trinity is a modelling approach, especially designed to support multi-actor environmental problem solving processes. ***Trinity*** is based on a cyclic model of problem solving. Focus is on three important stages of this cycle: analysis of the situation "as is"; construction of a script for taking action; and prediction of the improved situation "to be", expected to result from executing the script.

According to the ***Trinity*** approach, each of these three stages in principle is a *modelling process*. This results in three models: a descriptive model of the situation "as is"; a prescriptive model of the actions to be undertaken; and a predictive model of the situation "to be". In combination, they model a *perspective*: a body of knowledge that motivates and guides the process of intentionally turning a problem situation into a better situation.

To support the modelling of perspectives, the ***Trinity*** modelling language is designed. It consists of three model primitives (referring to environments, processes that change environments, and intentional actors); a set of connection constraints that enable to combine these model primitives, resulting in more complex models; and a set of abstraction and specification procedures that enable construction of models on several levels of detail.

In this paper, the conceptual fundaments of ***Trinity*** are presented, and parts of the ***Trinity*** modelling language are explained. The use of ***Trinity*** will be illustrated by means of a short example, concerning indoor environmental problems.

Keywords

environmental problems, problem solving process, actors, model, modelling, methodology, model-based support

1 INTRODUCTION

Trinity is a qualitative modelling methodology that is specifically designed to support problem solving processes in multi-actor situations. According to the ***Trinity*** point of view, a problem solving process is a *modelling process*. ***Trinity*** provides in a modelling language that supports several stages of this process. The outcome of the modelling process is a ***Trinity*** model: a model that motivates a plan for taking action, and guides implementation of this plan.
In this paper, first the premises and conceptual fundaments of ***Trinity*** are presented (section 2). After that, the ***Trinity*** approach as well as the role of models in this approach will be explained (section 3). In section 4, parts of the ***Trinity*** modelling language will be described: a language that enables to model complex systems of actors, acting in a shared physical environment. Section 5 presents a short example of a ***Trinity*** model, concerning indoor environmental problems. In section 6, the approach will be discussed and some conclusions will be drawn.

2 PREMISES

At first sight, "environmental problem solving processes" hardly refers to a homogeneous class of intentional activities. Instances of this class may differ for example with respect to the specific disciplines involved; the compartment where the undesired effect manifests itself (e.g. air, water, soil); the agent of concern (CO_2, heavy metals, noise); the actors involved; the geographical site; the scale (local, regional, national, international, global). At a closer look, however, it is possible to distinguish three "common denominators":

- Environmental problem solving processes in many cases are *multi-actor processes*;
- These actors act in a *shared environment* (which forms the "backbone" of an environmental problem context);
- Within problem solving processes *recurring patterns* can be recognised.

In this section, each of the three common denominators will be addressed separately. The ***Trinity*** approach integrates these three "common denominators": this will be worked out in section 3.

Multi-actor processes

Typically, in environmental problem solving processes a large *number of actors* is involved. These actors are either part of the problem context as such, or contribute knowledge to the problem solving process. They belong to many different societal groups (e.g. consumers, producers, governments) and scientific disciplines (e.g. natural sciences, technologies, human sciences). Apart from the number of actors involved, another important characteristic is the *complexity of the points of view* that motivate and guide the actions of these actors.
In figure 1, both attributes (*number of actors involved* and *complexity of points of view*) are used to construct a typology of environmental problem situations that focuses on knowledge distribution (for an elaboration of this typology see Diepenmaat, 1995). The letters A-D represent typical problem situations. The words in italics refer to typical problem solvers for these situations. Problem types and problem solvers are explained below.

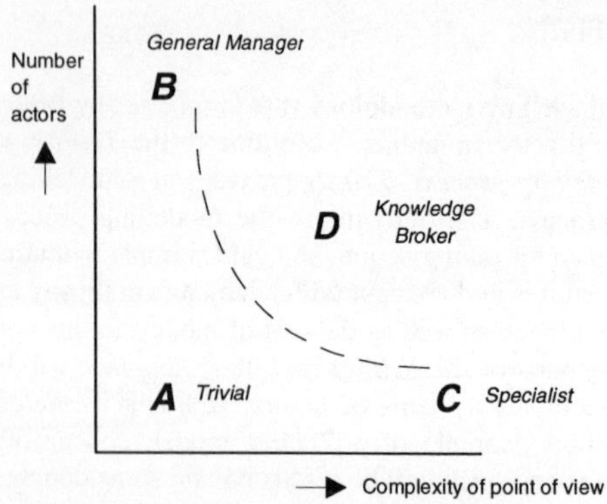

Figure 1 A problem typology on the basis of knowledge distribution.

A-type problems in principle are easy problems. Only one point of view (perspective) is relevant for solving the problem. Moreover, this point of view is simple.
In *B-type* problem situations, several simple points of view must be taken into account. This introduces a need for coherence and *co-ordination.* A typical problem solver is a *general manager*.
C-type problem situations require availability of a *specialist.* Note however that, although using the solution of a specialist may be easy, actually understanding his/her reasoning processes may turn out to be difficult.
D-type problem situations are worst case: they are characterised by a lack of coherence and organisation in a complex, multi-perspective interdependent context. Consequently, the key to a solution is a form of coherence between points of view, in which the original problem is solved, *and* no new insurmountable problems are introduced. A typical problem solver in *D-type* situations we call a *knowledge broker*.

The ***Trinity*** approach is directed at supporting *D-type* problem solving processes: the intended user is a knowledge broker. A knowledge broker is responsible for bringing about a solution to a *D-type* problem situation. Typical examples are policy makers and managers.
A knowledge broker is working in a problem context, involving many different points of view. These points of view can have a personal, disciplinary and/or cultural background. A commonly used strategy in this type of situations is to train people as *generalists.* Rather than *possessing* knowledge, however, the ability to *obtain, communicate and integrate* knowledge within reasonable limits of time and effort is a crucial requirement for knowledge brokerage activities. This implies an orientation towards knowledge *processes*, rather than towards possessing advance knowledge. In summary, a knowledge broker can be characterised as:

- a domain knowledge generalist;
- a knowledge process specialist; and
- oriented at knowledge, possessed by other actors involved.

An overview of the relation between problem types, methodological considerations and typical problem solvers is presented in table 1.

Table 1 Problem types, methodological principles and typical problem solvers (Diepenmaat, 1995)

Problem type	Methodological principles	Typical problem solver

A-type	Trivial	Trivial
B-type	Co-ordination	General Manager
C-type	Domain specific	Specialist
D-type	Knowledge modelling, Knowledge communication and Knowledge integration methods	Knowledge Broker

A shared environment

Different interpretations of environmental problem contexts exist (see for example de Groot, 1992, la Rivière, 1991). According to the ***Trinity*** approach, in multi-actor problem contexts three different, yet tightly related, domains must be distinguished (for a motivation see Diepenmaat, 1993). Real world phenomena are assumed to be complex constructs of processes and states in these three domains (hence ***Trinity***). The three domains are explained below:

- The **knowledge domain** encompasses the points of view of actors (e.g. a farmer's knowledge about how to cope with fungi on crops). The ABCD typology presented above is a knowledge domain typology.
- The **communication domain** encompasses all the ways that actors use to inform or influence other actors. Examples are a letter, a scientific article, a (part of a) conversation.
- The **physical domain** encompasses all aspects of our physical environment. Examples are an oil spill, a polluted river, a causal chain from emission to effect.

The points of view of the actors involved in a *D-type* situation are a "whole" because they are linked to each other in a *shared environment.* Points of view are bodies of knowledge, that motivate and guide the actions of these actors in their shared environment. The physical domain and the communication domain constitute the shared environment. In order to support multi-actor environmental problem solving processes, it is important to provide knowledge brokers with conceptual tools that support the process of obtaining an understanding of complex three-domain problem contexts. For this reason ***Trinity*** provides in a modelling language (section 4).

Recurring patterns: a stage-based approach

A problem solving process is a special case of an intentional activity. Intentional activities are purposeful activities: an actor acts according to some plan, and in the possession of expectations. The many reasons that may cause an actor to act intentionally were summarised very tersely by Wigg (1994): *"We avoid pain and seek pleasure".*

Many different attempts have been made to describe intentional activities as a generic sequence of simpler steps. Examples are the policy cycle, Checkland's (1990) experience-action cycle, the group communication cycle, Schön's (1983) spiral model of appreciation, action and reappreciation, the participative modelling cycle in policy analysis (Geurts and Vennix, 1989), the learning cycle of Kolb (1984). According to the ***Trinity*** approach, intentional activities can be structured as a generic sequence (cycle) of *six* stages:

1. *acknowledge problem;*
2. ***analyse the situation "as is";***
3. ***synthesise a script for taking action;***
4. ***predict the situation "to be", expected to result from executing this script;***
5. *execute script;*

6. *evaluate the resulting "to be" situation in terms of expectations.*

Stages 2-4 result in a *perspective*: a body of knowledge that an actor applies to an environment, and that motivates and guides his/her actions in this environment. Referring to the six stages of an intentional activity, perspectives can be thought of as a conglomerate structure of *three* mental models: a descriptive model of the situation "as is"; a prescriptive model (a script) for intervention in this situation "as is"; and a predictive model of the situation "to be", expected to result from actually executing the script. Distinction of these three parts is an important feature of the ***Trinity*** cycle, because in order to be able to act intentionally, an actor must possess *all three* of them: the difference between the model "as is" and the model "to be" motivates taking action, and the script guides this action. Note, that the sequence of six stages applies to widely differing activities, from global environmental policies to doing the shopping (although in routine activities stages 2-4 become trivial, perspectives are readily available).

A problem solving process is a process in which steps 2-4 require substantial cognitive effort: acquiring a perspective is not trivial. ***Trinity*** is directed at supporting steps 2-4 by means of constructing, discussing and exchanging models of perspectives. In many cases, several perspectives will be constructed: steps 2-4 will be executed several times. Stages 2-4 do not necessarily take place in this sequential order. In complex problem solving processes, the perspective is the outcome of a complex mixture of all three stages.

3 THE ***TRINITY*** APPROACH

The ***Trinity*** approach integrates the three common denominators of section 2: it enables to model systems of *several* intentional actors, acting in a *shared* environment, and the approach is structured according to the *generic, stage-based model* of intentional activities.

According to the ***Trinity*** approach, the processes of obtaining an understanding of a complex three-domain problem context (stage 2), formulating scripts to intervene in this context (stage 3) and predicting the outcome of actually executing the script (stage 4) are *modelling processes*. ***Trinity*** provides in a modelling language that guides and supports these processes. The outcome of the problem solving process is a qualitative model (or several models) that motivates and guides actions that are expected to result in an improved situation: the difference between the model "as is" and the model "to be" motivates taking action, and the script guides the implementation process (stage 5). Evaluation (stage 6) is supported by the models as well, as during this stage the actual situation "to be" is being compared with the predicted situation "to be". Stage 6 is where experiential learning takes place. This implies that, although ***Trinity*** models are constructed during stage 2-4, they support the *complete* intentional cycle of 6 stages.

The notion of "problem" is highly recursive: during a problem solving process, a stage may turn out to be problematic, which introduces a shift in problem context. A new recursive problem cycle is encountered, and ***Trinity*** may be applied again. A graphical summary of the ***Trinity*** approach is presented in figure 2.

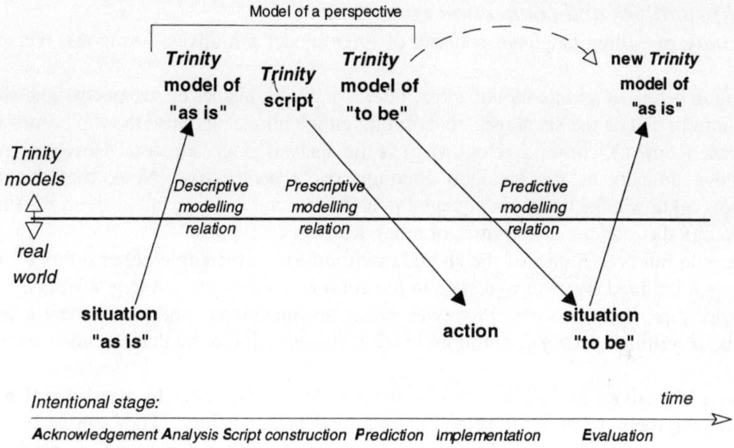

Figure 2 The ***Trinity*** approach.

4 THE *TRINITY* MODELLING LANGUAGE

Stages 2-4 of a multi-actor problem solving process are supported by modelling processes. In each of these stages, basically the same modelling language is used: the ***Trinity*** modelling language. This modelling language provides in model primitives and connection constraints that enable to represent complex three-domain perspectives. In addition, the ***Trinity*** modelling language provides in a set of abstraction and specification procedures that enable to model perspectives on different levels of abstraction. In this section, the model primitives and connection constraints will be presented. After that, abstraction and specification procedures will be globally described. Finally, the relation between the three partial models (model "as is", script, and model "to be"), in combination constituting a ***Trinity*** model of a perspective, will be clarified.

Model primitives and connection constraints

The ***Trinity*** modelling language consists of three model primitives: hexagons, rectangles and ellipses.

A **hexagon** refers to an intentional actor, interacting with his/her environment. The six corners can be interpreted as the six stages constituting an intentional activity: the top corner is stage 1 (acknowledgement), corner 2 (clockwise) is the analysis stage, etcetera. Hexagons model the knowledge domain, as the hexagon encompasses a perspective. Note, that this implies a recursion: a knowledge broker is engaged in an intentional activity, but at the same time he/she is modelling the intentional activities of other actors.

A **rectangle** refers to a part of the shared environment. A rectangle either refers to a physical state (e.g. a polluted river) or to a state in the communication domain (e.g. a letter).

An **ellipse** refers to a process. Processes either are intentional (they implement a script, e.g. someone is writing a letter) or autonomous (e.g. the emission of methane from a swamp).

Hexagons, rectangles and ellipses can be further elaborated, either by textual explanations, or by specifying them, resulting in more detailed models (see below). They can be connected by means of arrows, resulting in more complex models. Figure 3 presents a simple example. Arrows impose a partial ordering in time. They refer to causation: the model primitive at the tail of the arrow "causes" (is somehow responsible for) the model primitive at the head of the arrow. Five types of connections are allowed, each of them referring to another type of causation. The numbers next to the arrows in figure 3 refer to these types. They are explained below.

> ***Type 1*** *(rectangle → ellipse)*: a state causes a process.
>
> ***Type 2*** *(ellipse → rectangle)*: a process causes a state.
>
> ***Type 3*** *(rectangle → hexagon corner 1)*: a state is perceived by an actor, and causes this actor to apply (declare appropriate) a perspective.
>
> ***Type 4*** *(hexagon corner 5 → ellipse)*: an actor acts according to the applied perspective.
>
> ***Type 5*** *(rectangle → hexagon corner 6)*: an actor evaluates the results of his actions (compares expectations with the new situation).

Two types of ellipses (processes) can be distinguished: intentional processes (actions) and autonomous processes. The ellipses "Acts" and "Complains" in figure 3 are intentional processes, as they are caused by an intentional actor: they implement this actor's script. An example of an autonomous process is the ellipse "Causes" in figure 3.

Abstraction and specification procedures

Trinity abstraction and specification procedures enable to change the level of detail of a model. Parts of specific interest can be detailed, and less interesting parts can be abstracted. At this moment, we are finalising a library of abstraction and specification procedures. This library includes issues like handling loops and branching points in a model. Here, we will restrict ourselves to one example: the rectangle "environmental situation of concern" in figure 3 is specified to the "minimal environmental situation of concern", containing the essence of an environmental problem situation. The example makes clear that specification and abstraction are inverse operations. In addition, it exhibits the ground rule of ***Trinity*** abstractions and specifications: substitution of model parts is allowed if the "boundary" model primitives remain of the same type, and the connection constraints are not violated. Indeed, the boundary of the specified model in figure 3 consists of rectangles, as is the case for the abstract model.

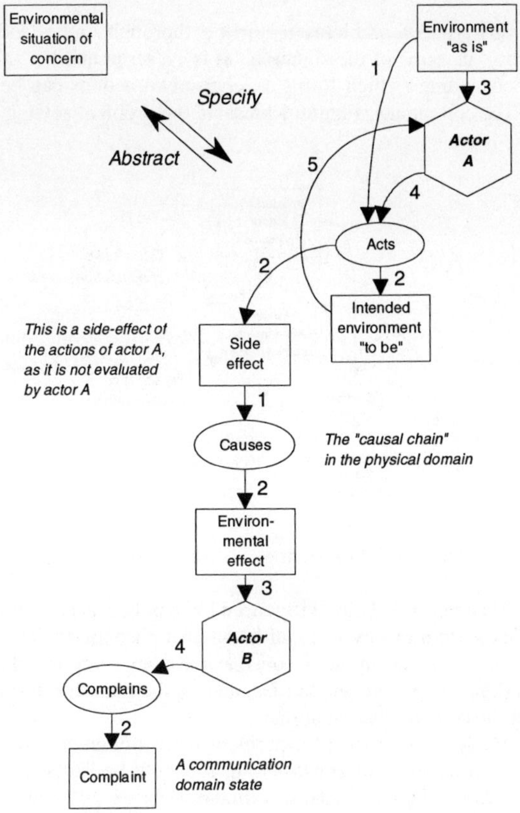

Figure 3 The minimal environmental situation of concern.

Models "as is", scripts and models "to be": three parts, one perspective

A model of a perspective consists of three partial models: a descriptive model of the situation "as is" (the situation without intervention); a prescriptive model of future actions; and a predictive model of the situation "to be". The relation between these three partial models is, that a script describes the transformation of the model of the "as is" situation to a model of the "to be" situation. The "to be" model no longer contains the undesired properties (i.e. the complaint).

Environmental problems are about emerging side effects in the physical environment, situated in a social context. All these elements are reflected by the specified model in figure 3. Therefore, we call this model the "minimal environmental situation of concern". "As is" models in general will be specifications of this generic model.

5 AN EXAMPLE: INDOOR ENVIRONMENTAL PROBLEMS

Due to the presence of specific agents (e.g. chemical agents like formaldehyde, ozone, biological agents like house dust mite) occupants of houses and offices may utter sub-clinical complaints. Solving this type of problems requires a thorough understanding of the indoor environment of concern, in terms of the situation "as is", a script and the situation "to be". On the basis of this understanding (which forms a perspective) actions can be taken in order to remove the cause of the complaints. Figure 4 presents the problem solving process in ***Trinity*** convention.

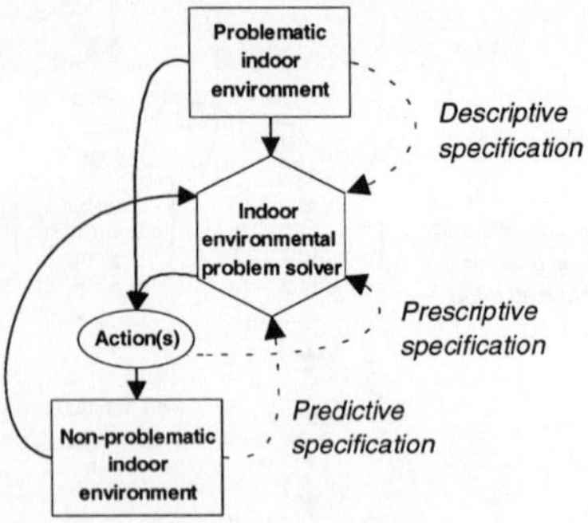

Figure 4 Indoor environmental problem solving.

Indoor environments are constantly being influenced by their occupants. Therefore, they will be modelled as a complex system of physical and intentional phenomena. The concept "actor" is the bridge between these two subsystems: on one hand an actor is able to formulate and possess perspectives (knowledge domain entities) and on the other hand an actor is able to act in the physical and/or the communication domain.

Figure 5 presents a generic "as is" model for problematic indoor environments. This model can be used as a first step in perspective construction: it should be further specified for specific problematic situations. For different agents, specified models are different as well. Consider for example rectangle 7 (Emission influencing circumstances). In a problem situation where chipboard is suspected to emit formaldehyde, typical emission influencing circumstances are temperature and humidity. It is therefore important to check whether the occupant (hexagon 4) likes a high temperature in the indoor environment. In situations where photo-copiers are suspected to emit ozone, on the other hand, the most important emission influencing circumstance is the intensity of use of the copier. In this situation, therefore, it is important to ask the occupant (hexagon 4) how often the copier is used. The generic model functions as a guide for further diagnosis of a problematic situation "as is". In a specific problem situation this results in more detailed models.

Obviously, scripts for taking action are closely related to the specific diagnosis. The general idea is, that interventions "branch off" the causal chain towards "Complaint" (rectangle 28 at the bottom), and result in a "to be" situation in which the complaint has disappeared. It is important to note that in principle *any* branching point "upstream" of the effect is a potential solution: removing the source (rectangle 8), minimising emission, maximising dispersion (ventilation), minimising external exposition, etcetera. This is an important benefit of models like figure 5: they are the result of co-operation between *several* experts, and ensure that a solution depends on an assessment and potential integration of *several* possibilities, rather than on the preferences of a coincidental specialist. The modelling process as well as the resulting model is *D-type*.

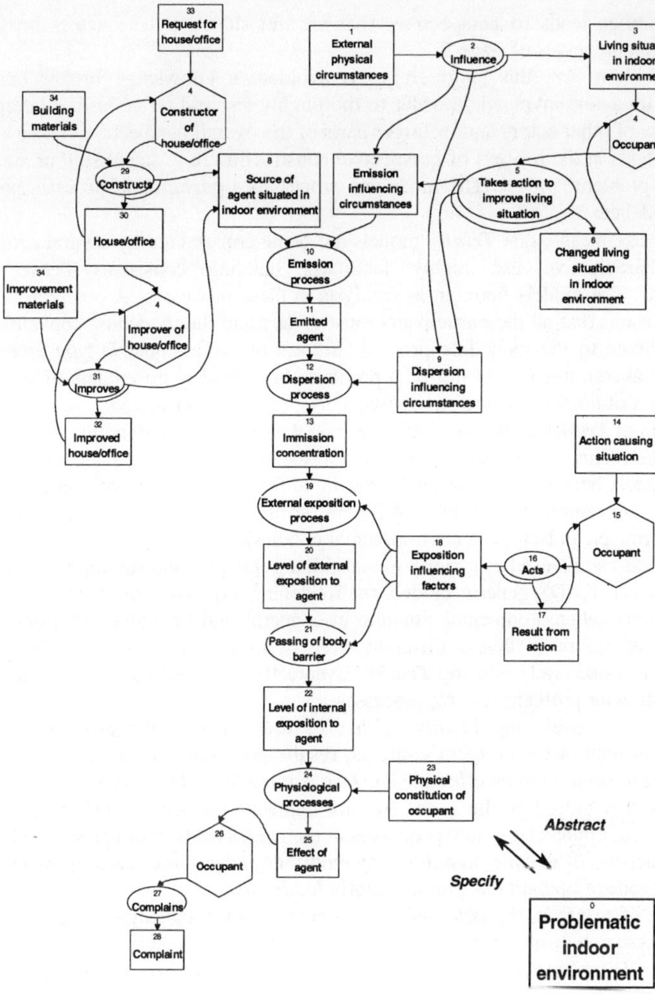

Figure 5 Generic descriptive specification of a problematic indoor environment.

6 DISCUSSION AND CONCLUSIONS

During the process of perspective construction (stages 2-4), three-fold ***Trinity*** models evolve from sloppy and ill-structured models to consistent models. Because of the fact that ***Trinity*** is founded upon a very generic model of intentional activities, its use can be integrated in multi-actor problem solving processes without disturbing common practice. The most important requirement is that the problem context is *D-type*.

The modelling process can take place in many different ways. Examples of different modes of using Trinity are *individual*, *hidden* and *participative* use.

Individual use means that an actor constructs models of multi-actor situations in isolation, by means of introspection. Although this mode of operation seems to be efficient and attractive at first sight, it often leads to perspectives that are not shared by the actors involved in the implementation processes of stage 5.

In case of *hidden use*, this shortcoming is avoided: a knowledge broker communicates intensively with actors involved, in order to thoroughly test and refine his/her interpretation of points of view of other actors and/or larger parts of the overall perspective. These actors either act as consultants in the process of perspective construction (e.g. scientists), or are participant in the actual problem context. The modelling process is a *background* process, performed by an expert modeller.

Participative use means, that ***Trinity*** models are being constructed by several actors in group processes. Participative use highly facilitates building consensus between differing "stakeholders". The models function as catalysts in these processes. A potential bottleneck in participative use is that all the participants must understand the modelling conventions, and be able to contribute to the modelling process. In case of professional *D-type* problem solvers (e.g. policy makers), this should not be a problem. In case of representants of societal actors, however, a mix of *hidden use* and *participative use* might be considered.

A strong point of ***Trinity*** is, that it enables to model complex three-domain contexts on several layers of detail. This is not only beneficial for the modelling process itself (as knowledge increases, models become more refined), but in addition models on different levels of detail may be used to communicate with different levels of decision making actors (cf. the subsidiarity principle in European environmental policies).

Abstract models can be used as generic models, that guide problem solving processes for more specific cases (cf. KADS generic models for modelling expertise, Breuker en van de Velde, 1994). The "minimal environmental situation of concern" and the indoor environmental model are examples of generic models at different levels of abstraction. A library of generic models will develop as a side effect of using ***Trinity***. Eventually, this library may function as a starting point for multi-actor problem solving processes.

The rationale for developing ***Trinity*** is that its use supports the process of perspective construction in multi-actor problem contexts, resulting in more coherent perspectives, which on their turn will result in more effective and efficient implementation processes. First exercises indicate, that this indeed is the case. At this moment, the use of ***Trinity*** is tested more extensively in environmental policy processes in the Netherlands. Examples of case studies are emission reductions of volatile organic compounds and reduction of packing materials. At this moment the mode of operation is predominantly *hidden use*.

In a new line of research, the potential of computer-aided knowledge brokerage is explored. This line of research is ultimately directed at designing and implementing a computer-based workbench for *D-type* problem solvers, that supports use of the ***Trinity*** approach (for example applying abstraction and specification procedures, consulting and extending the library of generic models).

Acknowledgements

The author wishes to express his thanks to Maarten Leupen, Jacqueline Cramer, Wim Harder, Bob Wielinga, Helmut Beckhuis and many others who have contributed to this research in several ways. This research is financially supported by the Dutch Ministry of Housing, Spatial Planning and the Environment.

7 REFERENCES

Checkland, P. (1990) Systems thinking, systems practice, Chapter 4. John Wiley and Sons, Chichester, England.

Diepenmaat, H.B. (1993) A comparison of models in the natural sciences, models in engineering and models of human expertise. TNO report IMW R 93/117.

Diepenmaat, H.B. (1995) A typology of environmental problems based on knowledge distribution: consequences for problem solving. TNO report IMW R 95/121.

Geurts, J.L.A. and Vennix, J.A.M. (eds) (1989) Explorations in policy analysis, the participative modelling cycle. Kerckebosch BV, Zeist, The Netherlands (in Dutch).

Groot, W.T. de (1992) Environmental science theory, concepts and methods in a one-world, problem-oriented paradigm. Elsevier, Amsterdam.

Kolb, D. (1984) Experiental learning: Experience as the source of learning and development. Prentice-Hall, Englewood Cliffs, N.J.

Rivière, J.M.W. la (1991) Co-operation between natural and social sciences in global change research: imperatives, realities, opportunities, *International Social Science Journal* **130**, 619-626.

Schön, D.A. (1983) The reflective practitioner, how professionals think in action. Basic Books, New York.

Breuker, J.A., Velde, W. van de (1994) The Common KADS library for expertise modelling. IOS Press, Amsterdam.

Wiig, K. (1993) Knowledge management foundations: thinking about thinking, how people and organizations create, represent and use knowledge. Schema Press, Arlington, Texas (p. 231).

8 BIOGRAPHY

Henk B. Diepenmaat (1962) studied chemistry at the University of Nijmegen, air hygiene and pollution at the Agricultural University of Wageningen, and knowledge engineering at CIBIT Utrecht / Middlesex University, Great-Britain. He is with TNO since 1987, at this moment as staff scientist. His main research activities concern the modelling of knowledge of multiple experts for knowledge transfer, knowledge integration and knowledge based systems development purposes, and methodological research with respect to model-based support for multi-actor problem solving processes, predominantly in environmental problem contexts.

12

EIS/GWM - An Integrated Automated Computer Platform for Regional Multimedia Environmental Audits

D. L. Toth
US EPA, Region III,
841 Chestnut Bldg., Philadelphia, PA 19107-4431,
tel. (215) 597-6688, FAX (215) 597-7906

Abstract

In this age of global budgetary constraints, environmental audits at the regulatory level must become efficient, effective and bold. EIS/GWM-"Environmental Impact System / Geomedia Waste Contaminant Migration" is an innovative PC computer platform that integrates data bases and contaminant migration simulation models, and offers an efficient way to establish an intelligent information system to perform a variety of remediation and health exposure studies and regulatory audits.

Keywords

Risk based environmental remediation, Regional environmental audits, automation

1 INTRODUCTION

The traditional way of performing environmental permit application reviews consist of appointing a Remedial Project Manager (RPM) per site and reviewing each site on an ad hoc basis. Little attention is paid on whether each site is influenced by, or affects other sites, or what would be an environmentally sound and an economically feasible approach. Consultants often times confuse client loyalty with personal gain, and end up proposing solutions which are both environmentally disastrous and economically undesirable. Regulatory Agencies must rise above partisan rhetoric and show in deed the way of a balanced environmental stewardship, all within the constraints of dwindling budgetary resources. The way out of this impasse is through the increased efficiency that the communications revolution has brought to bear. This paper illustrates the use of one software, EIS/GWM, in support of multimedia environmental audits.

Contaminant Migration and Remediation under EIS/GWM

Remediation and containment are the most common enabling technologies for the immediate control of the spread of contamination, and for the long-term remediation strategy, especially where natural degradation is allowed to play a constructive role.

Determination of the overall best remediation strategy is the objective of comprehensive Remedial Investigation/ Feasibility studies. These two complementary investigations represent respectively, the **diagnostic** and the **prognostic** aspects of the remediation process. The ***Remedial Investigation (diagnosis)*** is conducted concurrently with the ***Feasibility Study (prognosis)***, and emphasizes data collection and site characterization. The data collected during the diagnostic phase of the study are used to evaluate the existing state of the site, and to support the analysis and decision-making activities of the feasibility study, including the formulation of remedial alternatives. Comprehensive geomedia rehabilitation must include both aspects of the solution.

EIS/GWM is a prototype integrated environmental modeling platform, written for the MS Windows-3.x, to automate and support all aspects of the remediation problem. In this paper we summarize the sallient features of the platform, its architecture, and its operating environment. We also show pertinemt case studies where use of the platform not only streamlined the analysis phase, but also led to the most economical compliance solutions.

EIS/GWM offers a systematic, hierarchical approach, for the assessment of the relative impact of the various processes, by supporting and facilitating the analysis of the complex interactions. EIS/GWM manages the collection and use of all geographic, physical, and chemical data of a case study, entered in a graphical mode on-screen. It is used to analyze the fate and effect of chemicals in geological media. The results of this approach contribute to the better understanding of the physicochemical processes in underground flow and transport through the soil media; are useful in assisting Environmental Regulatory Agencies; and provide an integrated approach to the overall Risk Assessment of environmental contamination.

The EIS/GWM Platform offers unique capabilities to address both prognostic and diagnostic phases of remedial studies in a most efficient, reliable and cost-effective way. The basic activities of the Remedial Investigation and Feasibility Studies and their interdependency are illustrated in Figure 1. This overall methodology represents a systematic approach to achieving desired and reachable decontamination levels. The **EIS/GWM Platform** is the multimedia electronic integrator at the core of Figure 1, which includes:

1. Electronic data management to automate the flow of information, and streamline the overall effort and cost;
2. Health and safety planning to minimize risk of exposure; and,
3. Communications, to disseminate to the affected communities in the most efficient way the progress of the remedial studies using multimedia.

All these components are linked together in a uniform MS-Windows environment, providing task integration, the ultimate tool for every environmental auditor. These computer aided tools bring together and intelligently support various types of data used in environmental risk assessment studies, including: data collected from measuring devices in the field; information from private and public databases; scanned images; raster and vector satellite images; statistical data and risk maps.

2 CONTEXT OF THE CONTAMINANT MIGRATION PROBLEM

The EIS/GWM platform is based on a sound theoretical framework encompassing geostatistics, scale operators, and simulators. The formulation of the Remediation Problem is straightforward: determine the concentration 'C' of a given contaminant at all spatial locations of interest over time and propose remediating measures. What makes this determination difficult is the *large number* of: 1/ Interacting Geologic features (aquifers, geologic faults, aquitards, rock and soil formations), and 2/ Interacting Contaminant migration processes (biochemical processes). The platform excells in handling the different natural interactions that are the key elements of the groundwater remediation problem, by managing, accurately representing, and controling the number and the degree of these interactions under the requirements of a remediation investigation.

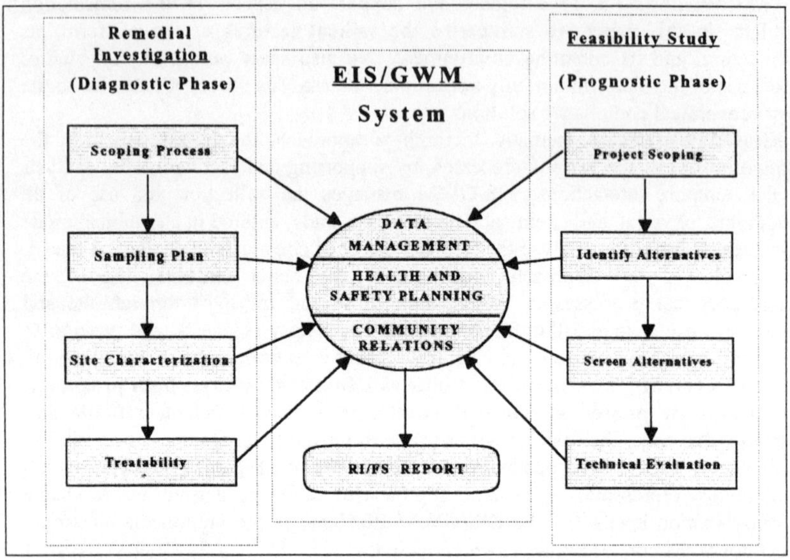

Figure 1. Interdependence of Diagnostic and Prognostic Phases of the Study.

The problem characteristics, in relation to the control volume and its impact to the overall solution, are illustrated in Figure 2. The control volume identifies the representative scale in which these interactions take place incorporated in a set of governing equations describing the balance of energy and mass over this confined domain. Different scales are necessary to accurately describe the evolution of different natural processes. What makes the EIS/GWM platform so effective is that the simulation framework is capable to *automatically* select the appropriate scale based on the information produced by the input **(diagnostic)** phase of the study. This information includes the constitutive parameters (properties) of the geomedia and the contaminants; and the influence of the ambient natural processes to the potential contaminant migration.

Clearly, the quality of this information has an impact on the simulation **(prognosis)** phase, where predictions are made on the fate and migration of different contaminants. This phase is critical to the success of the overall geostatistical effort and requires selection of a 3-D grid, characterizing the topology of the macro-element discretization of the natural domain. The inference engine is the processor which distributes the properties of the media throughout the grid. Finally, the integrator is at the core of the system, operating at the level of each macro-element and taking into account the effects of all the natural processes and their interactions described at the macro-scale and micro-scale levels.

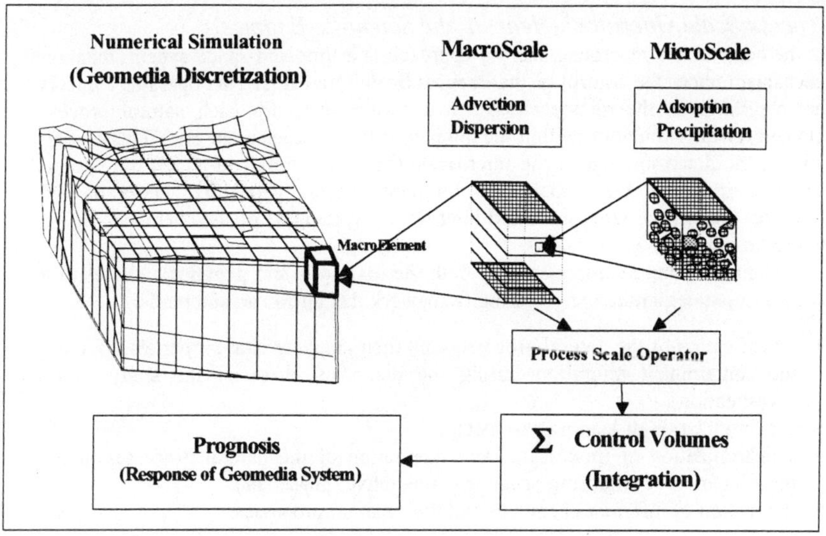

Figure 2. Soil media discretization and EIS/GWM scaling configurations.

3 OVERVIEW OF EIS/GWM AUTOMATED PROCEDURES

The automated procedures of the EIS/GWM platform are based on a macro-engineering framework, Figure 3. An informational network identifies the pertinent modeling features associated with each grid element and integrates all process interactions occurring within each macroelement using the appropriate balance equations. These equations reflect the interaction of the different natural processes that the user selected in the input phase. The contribution of each macroelement is therefore automatically incorporated in the informational network of the EIS/GWM platform at the location of each informational node in relation to the macroelements.

This computational framework allows the generation on-the-fly of the global set of matrix equations characterizing the simulation process. With the data properly entered and kriged, all that is left to do before activating a simulation is the selection of the initial conditions. Activation of the simulation and viewing of the results are easily performed with a click of the mouse. If the results are not satisfactory, several options are offered; the user can either change the simulation domain, alter the loading parameters, readjust the simulation grid, or redefine the initial conditions. The beauty of the platform is that these changes and alterations are built on-the-fly, <u>without</u> the need to reenter any of the fundamental data. The platform cleverly assists the user in each step, and keeps track of all the new parameters that enter the simulation. At the heart of all these automated operations is a process scale operator that synchronizes all interactions first at the level of the macroelement and second, at the global level of the entire site.

Process Scale Operator (Heart of the Scientific Engines)

At the heart of the macroengineering approach is a time-and-space synchronization mechanism under the control of the **Process Scale Operator**. This operator controls the integration at the microscale and macroscale levels for each natural process retained for the simulation within the domain of the macroelement grid. The starting point is the determination of the microscale (control volume) in which the natural processes influence the contaminant migration in the geomedia. These natural processes govern the fate (short and long term distribution) of the chemicals in the geomedia (air, soil, water, biota).

Managing all these components for both the **diagnosis** and **prognosis** phases of a study necessitates a macroengineering framework that allows an automatic:

- classification of the natural processes and their characteristic properties affecting the contaminant migration during the diagnosis phase of the study (site investigation),
- synchronization of loading parameters,
- synchronization of time scales and integration of the natural processes at the macroelement level during prognosis (predictive) phase, and
- interaction of different chemicals and their natural processes.

All these operations start within the confines of each macroelement and encompass the following procedural steps:

1/ Define "Fluxes" of different natural processes governing the simulation.
2/ Assess "Balance" of fluxes with source/sink terms that include 'Chemical Reactions' and 'Loading Terms'.
3/ Integrate 'Balance Equations' throughout the volume of each macroelement.

Typical fluxes as implemented in the various simulation modules of the EIS/GWM platform are also shown in Figure 4, along with the general description of the 'Balance equations'. Once these parameters are identified in conjunction with the needs of a particular simulation, the **Process Scale Operator** automatically integrates these equations throughout the volume of each macroelement.

Several schemes are available to perform this numerical integration throughout the volume of each macroelement accounting for all possible modeling features such as vertical and horizontal wells, slurry walls, geosynthetic or clay liners, drains, rivers, lakes and ponds. A Discrete Element scheme is implemented in the platform, fitting

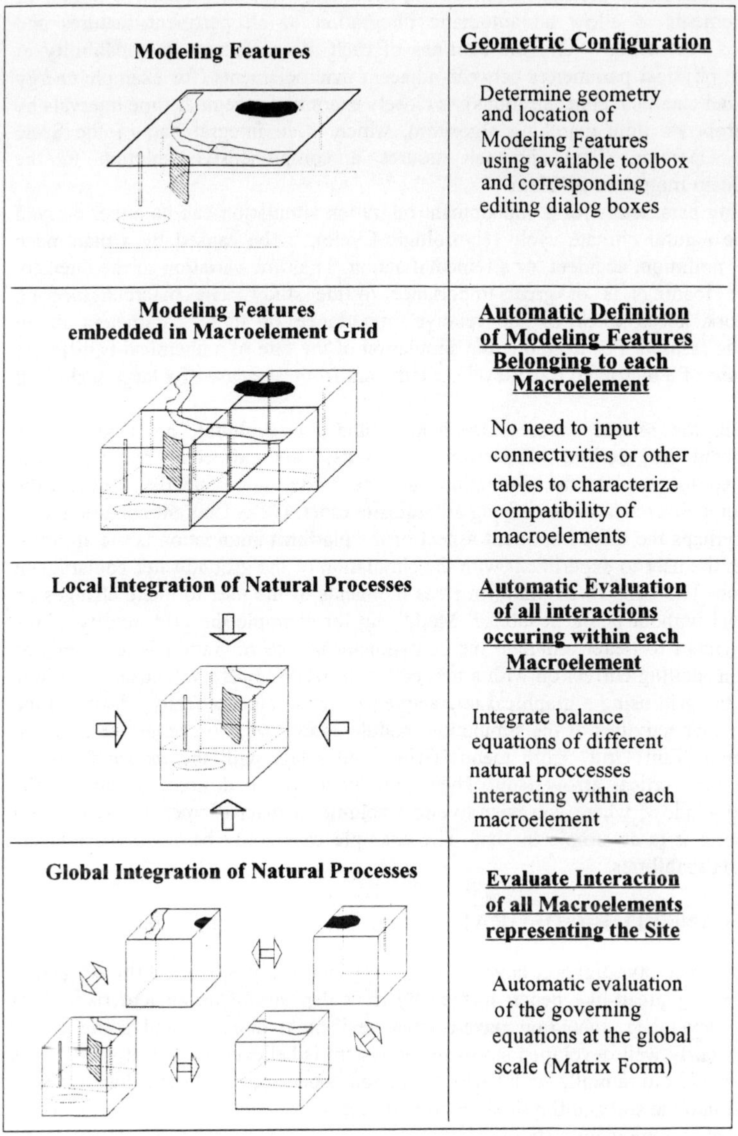

Figure 3. Built-in automated procedures under the EIS/GWM platform.

the requirements of the macroengineering approach, with an exhaustive library of macroelements to allow an automatic integration of all pertinent features and processes interacting within the confines of each macroelement. Compatibility of pertinent physical parameters between adjacent macroelements (for example energy fluxes, and chemical concentrations), is closely examined at regular time intervals by the appropriate time marching algorithm, which is an integral part of the Scale Control Operator. This approach assures a comprehensive solution to the contaminant migration problem.

Loading parameters for a contaminant migration simulation can be either derived from the natural climate cycle (Hydrologic Cycle), or be caused by a man-made activity, pollution, accident, or a remedial action. The time variation of the intensity of these loadings is of great importance to the study. The macroengineering framework keeps track of the relative importance of these parameters in an automatic fashion. For example, the simulation of the fate of a chemical is different in the case of a dissolved trace-level contaminant from the case of a large scale spill release.

Both, the release (source) mechanism and the evolution processes of the contaminant (advection, dispersion, reactions) are affected by the time incrementation used in the simulation; and the model automatically identifies the proper time discretization satisfying all requisite criteria (e.g. Courant conditions).

But perhaps the most important aspect of the platform automation is the ability it provides the user to experiment with the simulation of the groundwater contaminant migration. The EIS/GWM platform frees the hands of the user to make changes on his model without being penalized. Modifying for example the grid density of the macroelement to better simulate the contaminant source migration is as simple as making a spelling correction with a text editor. EIS/GWM offers the user the power to alter the grid using a graphical interactive mode that automatically adjusts on the spot the connectivities of the simulation module as follows. To change the grid you use option "Edit Grid" from menu "Grid", and select with the mouse from the toolbox the vertical arrow icon. Then, you move to the desired location in the working window, where you wish to add a column of macroelements, you click on the mouse: it is as simple as that. The example case study below showcases the platform capabilities.

4 ADVANCED BIODEGRADATION CASE STUDY

The site under investigation covers an area approximately 3,000 x 3,000 ft. and is seated on a plateau-like bench formed by river deposits of an ancient river. It is entirely developed, including paved areas, residential housing, and some grassy areas. A fairly well developed storm sewer system is believed to collect most of the local runoff from rainfall. An LNAPL (light non-aqueous phase liquid) fuel has been released into the soil and the shallow groundwater at the site near two oil tank farms. A detailed bioremediation study is initiated to predict the extent and migration of the spill shown in Figure 4. Different remediation scenarios are also examined.

There are two deeper aquifers at the site, but little information is available about them. However, the contamination is believed to be limited to the shallow (unconfined) aquifer, which is the subject of this investigation. The only available information on depth and thickness of the geologic formations is given for cross-sections A-A' and B-B', respectively in the general direction of flow and normal to it. In these cross-sections, the shallow aquifer is portrayed as limited to the sandy stratum which is underlain by clayey and sandy silt deposits. These deposits likely form an aquitard stratum separating the shallow aquifer from the Sunset aquifer. Samples collected from deeper pockets of clayey sand indicate that contamination may have reached these depths. Because the site is fairly well developed with paved areas and with a storm sewer system, it is believed that most of the recharge takes place upgradient of the site.

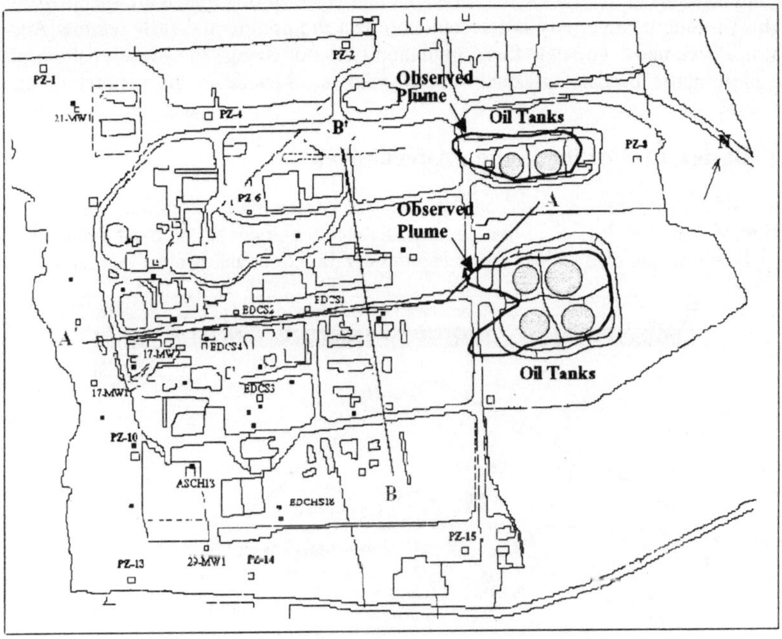

Figure 4. Site layout and cross section location.

The groundwater flow is generally in SouthWest direction. However, the "V" shaped delineation of the south contaminant plume necessitates a closer examination of the local flow regime west of the oil tanks. A previously "calibrated" flow model does not explain the bifurcation of the plume. In the present study we conduct a more refined investigation to verify the existence of a preferential flow path. In this instance the observed plume will be used as a tracer to explore the groundwater flow regime, assuming that advective conditions prevail at the site.

Ground water flow in the shallow zone is in the SW direction, with an average gradient of 5 ft per 100 horizontal feet. A local site investigation west of the main group of oil tanks revealed the existance of burried inactive conduits that could impose a preferential flow path, strong enough to alter the shape of the contaminant plume. The porosity of the aquifer has not been modified. The selected value of 0.25 is at the low end of values listed in the literature and seems adequate. The low value favors faster seepage velocities, therefore it is on the conservative side of the analysis. This argument alone is not sufficient to justify the selection, because ultimately one must show that the model used is capable of reproducing both observed piezometric heads and contaminant plumes.

This is where the different modeling tools of the EIS platform become handy. To model the preferential flow path created by the buried conduits a 20 ft. wide vertical trench (flow discontinuity) is selected from the toolbox and placed at the location of the old conduits as shown below. Then the properties of this trench are specified. A high hydraulic conductivity is selected to reflect the preferential flow regime. And that is all we need. The new flow simulation does not change the overall pattern of the piezometric heads. Only the velocity field is affectede in the vicinity of the conduits.

4.1 Design of Bioremediation Alternatives

The model calibration is based on the observed contaminant migration, spanning a period of one year. It is clear that an immediate containment action must be taken to stop the migration of BTEX in the downgradient direction, and protect key facilities.

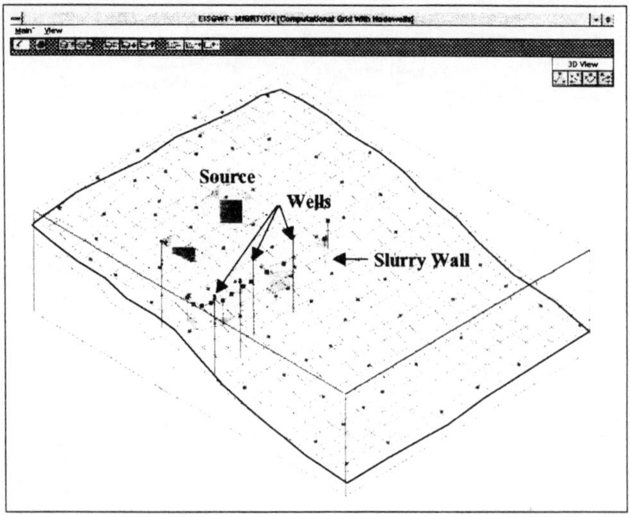

Figure 5. Slurry wall configuration.

One design alternative for containment is to build a slurry wall along the perimeter of these facilities. Then to further reduce the level of the BTEX concentrations, four injection wells are placed in front of the wall to reduce the contaminant concentrations. Oxygen with water are to be injected periodically in the wells to enhance natural biodegradation. The objective of the simulation is to check the feasibility of such design. The slurry wall is introduced in the model graphically through the use of slurry wall element in the toolbox. A similar approach is adopted to incorporate the injection wells, illustrated in Figure 5.

We implement this case following the steps introduced during the calibration of the model, starting with the flow module and continuing with the bioremediation simulation for 1000 days. The results of the simulation are conclusive. The shallow slurry wall is capable to contain the BTEX migration. After 800 days there are still some contaminant residuals behind the wall, but for the most part the level of the concentration has been reduced. Improvements can be made in optimizing the effectiveness of the pumps. What should be remembered is that a bioremediation model should be based on a rational well thought-out plan that includes:

- a clear definition of the goals of the study,
- an assessment of the quality of the site specific data, and an understanding of the limitations and quality of the results.

5 CONCLUDING REMARKS

In this article about the EIS/GWM platform and its use for environmental audits we have made the following points:

- Simulation models are necessary for ground water pollution remediation.
- Pumping tests provide local (intermediate scale) estimates of aquifer hydraulic properties and should not be confused with models.
- For proper use in predicting future system behavior, the model must be calibrated and verified. Calibration is the tuning of the model's parameters to match one set of observed conditions. Validation is the proof that the model predicts another set of conditions without need of further tuning.
- A good ground water biodegradation model is a suite of integrated simulation modules (flow between aquifer cells, recharge, pumping, ET, connection with surface waters, dispersion, chemical reactions, biodegradation) which render the calibration exercise rather arcane: an art form requiring substantial experience, reserved for the few.
- The EIS/GWM platform has made this art form accessible to everyone.
- This is the only platform suporting multiple plume 3-Dimensional simulations.

In this approach, not only does the regulator find distilled a multi-year experience in modeling and calibration, but he also finds the tools to use this experience tangibly in the form of the EIS/GWM platform.

6 REFERENCES

1. MicroEngineering, 1994. "BIOREM-3D, Contaminant Biodegradation Models Under the EIS/GWM Platform," Report No. Mi-94-M009.

2. MicroEngineering, 1994. The EIS/GWM Intgrated Computer Platform: "Environmental Impact System for Contaminant Migration Simulataions and Risk Assessment," Report No. Mi-94-M001.

PART FOUR

Environmental Management

13

Computer-supported environmental management – why simulation methods are useful

L. M. Hilty
Department of Informatics, University of Hamburg
Vogt-Koelln-Str. 30, 22527 Hamburg, Germany
Tel: ++49/40/54715-309, Fax: -311, hilty@informatik.uni-hamburg.de

Abstract

Two approaches to use computer simulation methods for ecological assessment are presented: The "Material Flow Network" approach based on the well-known formalism of Petri Nets and the "Eco-Logistics" approach based on the integration of simulation and geographic information systems. These two approaches allow investigation of the dynamic behavior of complex systems in the conflict between economy and ecology. They can therefore serve as a central part of decision support systems for environmental management, supporting the simulation of "what if"-scenarios and aiding the user in finding a strategy with high ecological efficiency.

Keywords

Environmental Management System, Life Cycle Analysis, Simulation Modeling

1 BACKGROUND AND TERMINOLOGY

Environmental management needs to be supported by computer-aided environmental information and management (CAEM) systems. Although comprehensive software systems for this purpose are not yet available, there are some programs that support the basic task in environmental management: environmental assessment.

However, the methods and software tools for environmental assessment developed so far are backward-oriented, static, and do not account for causal mechanisms in the complex economical and ecological systems under study.

The need for environmental management systems in industrial production is a recent challenge to information technology, and has stimulated a broad range of activities in Applied Computer Science particulary in Europe (Hilty, 1994a, 1995). An *environmental management system* is the overall framework for the actions that an enterprise takes to manage its effects on the environment (Little, 1993). Regulations such as the Environmental Management Standard BS7750 (British Standard Institution, 1992) and the EC "Eco-Management and Audit Scheme" (Commission of the EC, 1993) are intended to standardize this framework. Information systems designed to support such actions or the overall framework are sometimes called Environmental Management Information Systems (EMIS), Industrial Environmental Information Systems or Computer-aided Environmental information and Management (CAEM) systems (Hilty, 1995), depending on the context in which the respective systems were designed. The terminology in this new field is far from being consistent.

One specific approach to such systems is to extend existing production planning and control (PPC) systems in such a way as to help to reduce emissions into air, water or soil. Systems emerging from this approach are also known as Environmentally-oriented PPC (EPPC) systems (Haasis, 1992). Considerable emission reductions ar\e possible when such systems are implemented in industrial production plants. Moreover, the idea of integrating information systems for production and recycling leads to Production and Recycling Planning and Control (PRPC) systems (Rautenstrauch, 1994).

Other specific approaches emerge from the field of so-called "ecobalances". This term refers to studies or methods that are used to investigate the mass and energy flows in a given system, assess their ecological impacts, and (e)valuate them in the framework of ecological valuation criteria. There are at least two different ideas behind the term "ecobalance":

1. The ecological effects of any system can be described on the physical level in terms of mass and energy flow through (and transformation caused by) the system. Therefore, the physical inputs and outputs of the system can be put on the two sides of a balance sheet (see figure 1). According to the first law of thermodynamics, input and output must be balanced if measured in physical units. The transformation caused by the system can be described as an increase of entropy, according to the second law of thermodynamics (entropy law).
2. Environmental assessment is regarded as an extension of traditional business accounting, assuming that environmental effects should be recorded and analysed with the same systematics and accuracy as monetary transactions are. This idea traces back to the Swiss engineer Müller-Wenk, who invented a system of ecological accounting based on a non-monetary (and non-physical) measuring unit for environmental effects, expressing the ecological scarcity of limited resources, including limited capacities of regeneration processes.

In Europe, a broad range of methods for ecological assessment are subsumed under the notion of ecobalance today. An important group is formed by ecobalances for products, better known as Life Cycle Assessments (LCA). The goal is to assess the environmental impacts of a product "from cradle to grave", i.e., from the extraction of raw materials up to waste disposal. There is a variety of LCA methods (Weidema, 1993).

A second group of ecobalances considers one company or one plant of a company. It can be viewed as a special case of the earlier concept of "social accounting", which aimed to account for social impacts of a company's activities in general ("put people on your balance sheet"). This approach may have failed because of its comprehensive demand. However, putting the *environmental* impacts of a company onto a balance sheet seems to be the easier (because more specific) task, although it is complex enough and may not even be characterized as a well-defined problem.

One of the first approaches to ecobalances for companies was Müller-Wenk's "ecological accounting" developed in the seventies. It is based on the concept of ecological scarcity, which allows for the quantification of all types of environmental impacts on a single scale. This approach was later developed into the method of eco-balances recommended by the Swiss Federal Office of Environment, Forestry and Landscape.

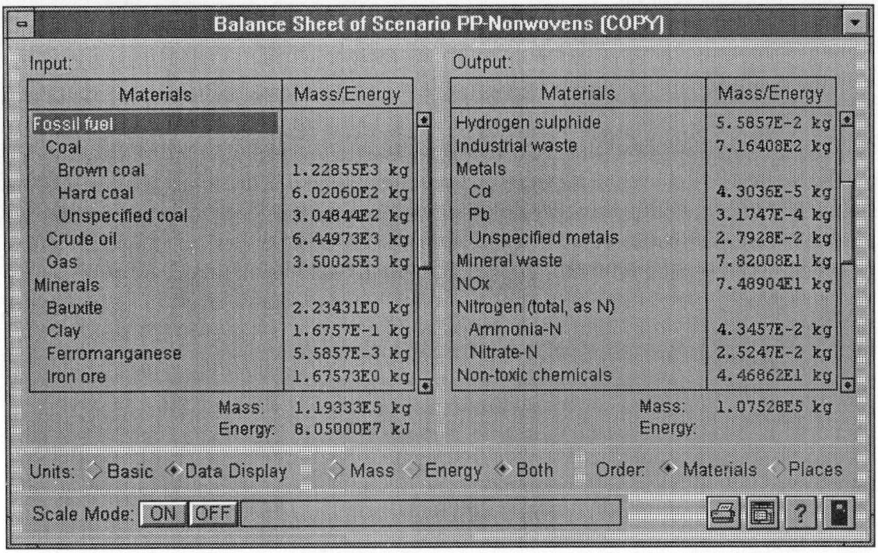

Figure 1 Small section of an ecological balance sheet for polypropylene (PP) production. Mass input and output are approximately balanced, whereas energy is not. (The reason for the latter is that waste heat was not measured for this study.) This figure is a screen copy from the *Umberto* software.

2 MOTIVATION FOR ENVIRONMENTAL MANAGEMENT

Environmental assessments of companies, products, or production processes are the starting point of environmental management. The result of an environmental assessment is a detailed documentation of environmental effects. These are changes to the environment caused by an activity. They can be divided into resource consumption (raw material and energy consumption), emissions or discharges (gaseous, airborne, liquid, dissolved or solid waste), and welfare effects on humans and possibly other organisms (including health effects).

Based on the information gained in a basic environmental assessment, a company can check legal compliance, identify the most significant environmental effects, and define an environmental policy. Further steps to environmental management are the initiation of more detailed environmental assessments of selected aspects, the arrangement of independent environmental audits, and finally the introduction of an environmental management system (Little, 1993, see figure 2).

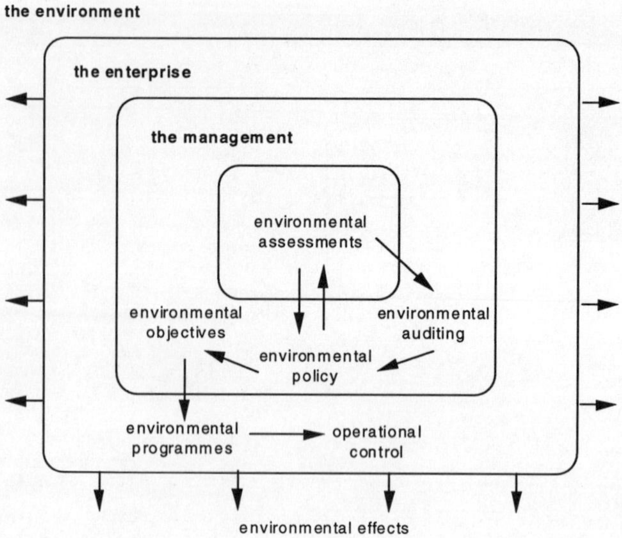

Figure 2 Elements of an environmental management system (following Little, 1993).

The main motivation for environmental management is the *external demand for environment-related information* about the enterprise and its products:

- Consumers increasingly consider ecological criteria when choosing a product or a service. This results in a demand for product- and company-related environmental information.
- Suppliers and buyers, in expectation of a long-term business relationship, need environmental information of their partner in order to minimize the risk of getting involved in environmental problems and scandals.
- Investors (shareholders and creditors) often consider ecological aspects. Banks, for instance, take the potential contamination of a site into consideration when granting a mortgage. Environmentally oriented investors choose the least ecologically damaging company if the capital return is otherwise equal.
- The insurances industry has to identify and evaluate the increasing liability risks resulting from environmental hazards.

3 SOFTWARE SUPPORT FOR ENVIRONMENTAL ASSESSMENT

To classify computer programs for environmental assessment, we differentiate between the following three levels of an environmental assessment, also known as the three steps of "ecobalancing":

- *inventory assessment:* quantification of all (relevant) mass and energy flows caused by the system under study, including energy and raw material input and the related output in solid, liquid, or gaseous form; description and quantification of other forms of interaction with the environment (e.g. land utilization).
- *environmental impact assessment:* description and quantification of the environmental effects caused by the interactions determined in the invertory phase.
- *evaluation:* evaluation and presentation of the results of the preceeding phases to make them suitable for decision support. In this phase, priorities or weights are given to parameters of environmental effects, and different kinds of impacts are aggregated. Because this involves *values* concerning the relative importance of the parameters (e.g., health effects versus energy consumption), this phase is also called the *valuation* phase.

In some publications, another phase preceding the inventory assessment (goal definition, scoping/screening) is included (Fava, 1993; Umweltbundesamt, 1992). It comprises, among other issues, the definition of the system boundary, the choice of methods for the succeeding phases, and defining a strategy for data collection. At present, there seems to be no systematic approach to this first phase of environmental assessment nor, consequently, appropriate software support.[1]

The available software systems usually focus on one of the three steps listed above. They can therefore be classified into *inventory-*, *impact-* and *(e)valuation*-oriented systems. Most systems mentioned here are described in more detail either in the textbook "Product Life Cycle Assessment" (Weidema, 1993) or in the proceedings of the conference "Informatics for Environmental Protection" (Hilty, 1994a), volume II.

3.1 Inventory-oriented Systems

The most basic step of ecological assessment is collecting reliable data and integrating it into a model of mass and energy flow which is a consistent and valid representation of the system under study.

One of the first software systems developed in this field was the *LCA inventory tool* (Chalmers Industrieteknik, Sweden), which supports modeling product life cycles with directed graphs. A database with the emission data of the most important production, transport and energy generation processes is provided.

The *Umberto* system (Institute for Environmental Informatics, Hamburg, in cooperation with the Institute for Energy and Environmental Research, Heidelberg, Germany) supports the modeling of mass and energy flow systems based on so-called material flow nets, an interpretation of Petri Nets. Ecobalances of companies, products, production and disposal processes on different system levels can be treated the same way. An expandable library with predefined modules for energy production, raw material production, transport and waste disposal is available (*Umberto* is discussed in more detail in section 4).

Further inventory-oriented systems are *TEA* (Tools for Ecological Analysis, Ecobilan, France), *IDEA* (International Database for Ecoprofile Analysis, Technical Centre of Finland), and *TEMIS* (Total Emission Model for Integrated Systems, Öko-Institut Darmstadt, Germany).

1 This may change in future when the ISO 14000 series is established.

3.2 Impact-oriented Systems

For an estimate of ecological impacts, chemical substance data and their behaviour in the environment are needed.

The program *SimaPro2* (Pré, The Netherlands) was conceived to support ecological product development. With the use of a database for materials and production processes, the direct and indirect environmental effects of a proposed product variant can be assessed.

Furthermore, there are several on-line chemical databases offered worldwide which document the environmental effects of chemical substances, e.g., the *IRPTC* (International Register of Potentially Toxic Chemicals, Health and Welfare, Canada).

3.3 (E)valuation-oriented Systems

The program *EPS* (Environmental Priority Strategies, Swedish Environmental Research Institute) supports the evaluation of resource consumption and emissions by quantifying the environmental burdens in monetary units. Other valuation concepts (like ecological scarcity) can be considered as well.

EXCEPT (IBM Germany, in cooperation with the Technical University of Hamburg) is a knowledge-based system which supports complex, rule-based ecological valuation methods. Here, the most important aspect is the documentation and explanation of valuation methods and the comparison of alternative assessments (Weiland, 1993).

4 MODELING SYSTEMS OF MASS AND ENERGY FLOW WITH PETRI NETS

Practical experience gained in large ecobalance projects lead to the development of the software tool *Umberto* (Schmidt, 1994. This tool is designed and implemented on the basis of Petri Net theory. Petri Nets have already proven to be useful in modeling systems of information flow; they are exploited in *Umberto* to support comprehensive and consistent modeling of mass and energy flow systems such as production systems or product life cycles.

Usually, life-cycle assessments are calculated by means of an extended matrix calculation, which causes the method to appear less transparent and more difficult to handle than it should be. These problems are overcome by means of the Petri Net approach. This formalism is suited for modeling systems in which flows of information, matter or energy play a central role. The systems are represented by graphs with two sorts of nodes which must strictly alternate: so-called *transitions,* representing transformation processes, and *places,* which represent, for instance, depots.

Net theory was combined with concepts of ecological accounting by Möller (1994) and led to a new type (or interpretation) of the Petri Net formalism: material flow networks. The pilot application of this approach and the *Umberto* software was an international LCA project in which data from more than 90 European companies were collected, structured and integrated into an LCA tree for disposable diapers. Figure 3 shows a small portion of the whole network, which would cover more than one square meter if printed on paper in the same size and aggregation level.

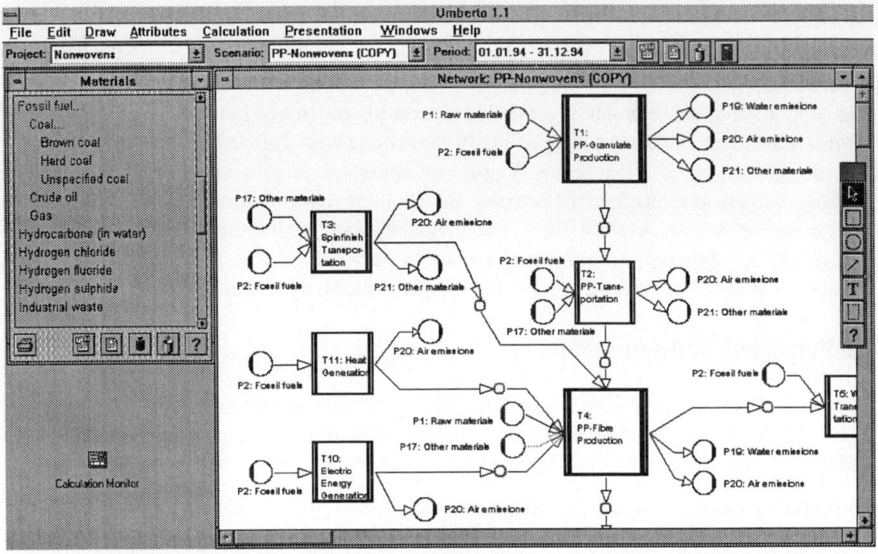

Figure 3 Small section of the network representirng the product life cycle of polypropylene (PP) nonwovens from raw materials to waste treatment. The window entitled 'Materials' shows a part of the project-specific materials classification. This figure is a screen copy from the *Umberto* software.

The following table shows how real systems are mapped to the Petri Net formalism and its graphical representation.

real system	Petri Net (conceptual model)	graphical representation	comment
process (production, transport, consumption/use, waste treatment)	transition (specified by an arbitrary mathematical function definition)	rectangle	The function defines the input/output relation.
storage (any place where material is located before or after being processed)	place	circle	One place can accommodate different materials.
material flow (flow from a place into a process or from a process to a place)	arc	arrow	Different materials can flow "through" one arc.

Different parts of the network can subsequently be combined to form a new transition. This makes it possible to build up hierarchical network structures. Moreover, it is possible to extract an input/output balance for the environmentally relevant parameters at any point of the network.

The Petri Net approach has considerable advantages for complex LCA projects. Mainly,

- it helps to model systems of mass and energy flow in a flexible and mathematically consistent manner;
- it supports interactive graphical modeling in both top-down and bottom-up directions (stepwise refinement of models and modeling by combining basic building blocks, respectively);
- it allows the use of a "spreading activation" approach to network calculation, so that a partial numerical evaluation of a network model or any model fragment can take place at any time; inconsistencies detected due to data redundancy are fed back to the user immediately;
- it allows an algorithm of truth maintenance to be used, so that recalculation of inferred values – after the user has altered a value – is performed only if necessary.

4.1 Petri Nets and simulation

If an industrial production system or product life cycle is once modelled in the form of a Petri Net, it is possible to simulate its dynamic behaviour. For decision support in environmental management, especially the following applications of simulation seem promising:

- comparing the environmental effects of planned product or process alternatives,
- assessing the consequences of introducing new materials into a product life cycle or production systems, especially under the aspect of waste treatment,
- planning and controlling integrated recycling and production processes under uncertain assumptions about return rates and quality of used products.

5 ECO-LOGISTICS AND THE "MOBILE" PROJECT

The second approach to apply simulation methods in environmental management to be discussed here requires again some preliminary remarks on terminology.

The concept of logistics, which originates from the military context, has been generalized to methods for coodinating people and/or goods in space and time in order to maximize the benefits and to minimize the cost. In economic terms, production factors (people, machines, material...) have to be combined for the purpose of production, and people and goods must come together for the purpose of consumption. The aim of logistics is an optimal design and control of systems of physical or information flow.

This general concept can further be extended to eco-logistics (Hilty, 1994c) by considering ecological scarcity of resources and environmental pollution in addition to monetary costs. This is important because the ecological impacts of the traffic sector are not adequatly indicated by the monetary cost of transportation.

Some simulation models have been developed for specific problems in this field, e.g. for the environmental impacts of "just in time" delivery and production strategies (Hilty, 1994b) and for commuter traffic in crowded areas (Hilty, 1994d).

The research project "Mobile" ("**Mo**del **b**ase for an **i**ntegrative view of **l**ogistics and **e**nvironment") at the University of Hamburg has the objective to develop an object-oriented model base system for eco-logistics. The model base is a growing collection of simulation model components that can be easily combined into problem-specific models for a broad range of applications.

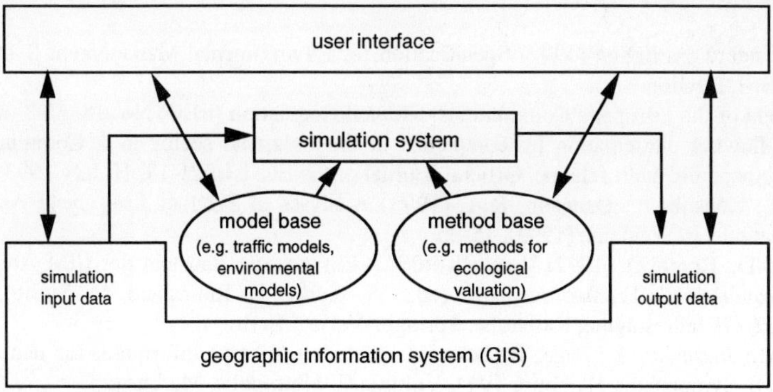

Figure 4 Architecture of the model base system developed in the Mobile project. This interdisciplinary project is supported by the *Deutsche Volkswagen-Stiftung*.

This approach differs from the Petri Net approach discussed in section 5 in the following respects:

- All processes can be located in geographic space, whereas Petri Nets have no such concept of space.
- Processes can be attributed to actors, and actors can change their location according to individual strategies.
- The combination of discrete-event and continuous model components is possible, which is necessary, e.g., when micro- and macroscopic traffic models are to be integrated or when discrete event emission models and continuous atmospheric dispersion models are coupled.

6 CONCLUSION AND OUTLOOK

There is a variety of approaches to support environmental assessments by sophisticated software systems, emerging from the field of "ecobalances". These approaches usually rely on *static* models of the system under investigation. However, it is necessary for environmental managment to consider the *dynamic* behaviour of systems, e.g. the interactions of production, consumption and recycling processes. This requires the use of computer simulation methods for decision support in environmental management.

Two approaches already try to meet this requirement. The material flow network approach implemented in the *Umberto* Software and the modelling and simulation approach to ecologistics developed in the *Mobile* project.

However, the environmental assessment of past or future decisions is only the basic part of environmental management. In the long run, integral Computer Aided Environmental Management (CAEM) systems will have to be developed. The overall framework for integral CAEM systems will be set by standards such as the ISO 14000 standards family, which is expected to play a similar role for environmental managment as the ISO 9000 series did for quality management.

7 REFERENCES

British Standard Institution (1992) Specification for Environmental Management Systems. BS 7750. BSI, London.

Commission of the European Communities: Council Regulation (EEC) No. 1836/93 of 29 June 1993 allowing participation by companies in the industrial sector in a Community eco-management and audit scheme. Official Journal of the EC, L168/1-18, 10 July 1993.

Fava, J.A.; Consoli, F.; Denoson, R.A. (1993) Analyses of Product Life Cycle Assessment Applications, in Weidema (1993), 15-26

Haasis, H.-D.; Rentz, O. (1992) "Umwelt-PPS" – Ein weiterer Baustein der CIM-Architektur? in *Information als Produktionsfaktor* (eds. W. Görke, H. Rininsland, M. Syrbe), Proceedings 22. GI-Jahrestagung Karlsruhe, Springer-Verlag, Berlin.

Hilty, L.M.; Jaeschke, A.; Page, B; Schwabl, A. (eds.) (1994a) Informatik für den Umweltschutz, 8. Symposium, Hamburg 1994. Volume II, Metropolis, Marburg.

Hilty, L.M.; Page, B.; Martinssen, D. (1994b) Designing a Simulation Tool for the Environmental Assessment of Logistical Systems and Strategies. In *Computer Support for Environmental Impact Assessment* (eds. G. Guariso, B. Page), IFIP Transactions B-16, North-Holland, Amsterdam .

Hilty, L.M.; Weiland, U. (1994c) Sustainable Cities – Opportunities and Risks of Information Technology. In *Proceedings of the 13th World Computer Congress* (eds. K. Brunnstein, E. Raubold), Volume II, Elsevier, Amsterdam, 613-18.

Hilty, L.M. (1994d) Ökologische Bewertung von Verkehrs- und Logistiksystemen – Ökobilanzen und Computersimulation. Hochschule St.Gallen, IWÖ-Beitrag Nr. 18, St.Gallen.

Hilty, L.M. (1995) Information systems for industrial environmental management, in *Environmental Informatics – Methodology and Applications of Environmental Information Processing* (ed. N. Avouris), Kluwer Academic Publishers, Dordrecht.

Little, B.; Weidema, B.P. (1993) Environmental Management Systems, in Weidema (1993), 27-38.

Möller, A. (1994) Stoffstromnetze. in Hilty (1994a), Volume II, 223-30.

Rautenstrauch, C. (1994) Integrating Information Systems for Production and Recycling, in Hilty (1994a), Volume II, 183-90.

Schmidt, M.; Giegrich, J.; Hilty, L. M. (1994) Experiences with ecobalances and the development of an interactive software tool. in Hilty (1994a), Volume II, 101-08.

Umweltbundesamt (1992): Ökobilanzen für Produkte. Bedeutung – Sachstand – Perspektiven. UBA-Texte 38/92, Berlin.

Weidema, B.P. (ed.) (1993) Environmental Assessment of Products. UETP-EEE – The Finnish Association of Graduate Engineers, Helsinki.

Weiland, U.; Pietsch, J.; Hübner, M. (1993) The EXCEPT-Project: Expert System for Computer-Aided Environmental Planning Tasks, in Proc. IFIP Working conference on Computer Support for Environmental Impact Assessment (CSEIA) 1993, Como, 358-67.

8 BIOGRAPHY

The author was born in Switzerland, took his doctor´s degree in Computer Science (Informatics) at the University of Hamburg and worked in various research projects in the field of environmental information processing. At present, he is in charge of the MOBILE project.

14

Quality Management as Basis of Environmental Management Systems for Industrial Plants

P.H. OSANNA *), D. PROSTREDNIK *), A. OSANNA **)

*) Department for Interchangeable Manufacturing and Industrial Metrology (Austauschbau und Messtechnik), Institute of Production Engineering, Vienna University of Technology - TUW, Karlsplatz 13/3113, A-1040 Wien, Austria.

**) Department of Physics, State University of New York at Stony Brook, Stony Brook, NY 11790-3800, USA.

Abstract

Quality management systems according to International Standards of the ISO 9000 series are a good basis for building environmental management practice. This offers the opportunity to apply well known systems and well known tools in a modified form to solve a company's environmental problems. Additionally, this will put the environmental question far beyond national boarders maybe throughout the world.

Legislation, consumers and environmental organisations put increasingly pressure on companies to improve their environmental performance. Therefore in various countries adequate management of environmental issues is becoming indespensible as a means to keep a companies competitive position.

Keywords

environmental management systems, quality management systems, small and medium sized enterprises

1 WHAT IS OUR POINT OF VIEW ?

Quality management based on ISO 9000 is a very advanced and efficient management system in various countries all over the world. The know-how and experience in quality management derives of the early years of twentieth century and since that time developed continously. Today a well developed quality management system is a key factor of every company's success. We find quality management systems in all kinds of industrial plants and economical structures and also other organisations according to (Osanna, 1991).

Legislation, consumers and environmental organisations are increasingly putting pressure on companies to improve their environmental performance. Thus, in an also increasing number of countries adequate management of environmental issues is becoming indispensable as a means to keep a company's competitive position.

2 CAN WE INTERCORRELATE ENVIRONMENTAL AND QUALITY ISSUES ?

Besides controlling the quality of products the management system on the basis of the series of International Standards EN ISO 9000 especially (EN ISO 9000-1, 1994) and (EN ISO 9004-1, 1994) can build in a similar way the basic system for handling the interactions between a company and the needs of ecology. The starting point for a comparison of systems must be to find the joint characteristics of quality and environmental issues.

An obvious possibility to combine quality and environmental issues is given by the definition of quality. According to the term standard (EN ISO 8402, 1994) quality is defined as the entity of features of a unit referring to the appropriateness to fulfill determined and premised requirements. Nowadays in industrialized countries it is widely agreed that the environmental compatibility of a product is a quality feature. But it is not a customer-specific, it is a more or less society-specific quality feature. The only question arising in this context is: How much are customers willing to pay for this "society-specific quality feature" environmental compatibility?

There is another reason for the link between environmental and quality management. The quality management system according to the EN ISO 9000 series was developed to satisfy costumer demands for quality. But the management principles are generally valid and can be used to solve and manage other society interests. In the same way, how technical specifications for quality are separated from the management system for quality, the technical specifications in these areas have to be developed separately too.

This separation of management system standard and different environmental standards - which depend on a lot of heterogeneous regulating actions in various countries and industrial branches - is an essential boundary condition.

These general ideas have been taken as a basis for the development of an environmental management system by the British Standards Institute as the standard (BS 7750, 1992) on the roots of the EN ISO 9000 series.

3 WHAT IS THE EUROPEAN POINT OF VIEW ?

At present there exist two different system approaches in Europe.

On the one hand the EC in general focusses systems on auditing of internal management systems as well as the publication of the results and a verification on this bases.

On the other hand according to (BS 7750, 1992) an all including management system should be developed and implemented which later on should be ready for audits. At present this approach is often entitled as the "ultimate standard". In reality it is more than just the sum of the two parts "quality" and "environmental management", it is a compound of both perspectives by taking the largest synergetic effects possible.

On this basis new proposals for international standards give general ideas for the principles of environmental management systems according to (ISO 14000, 1994) and guidance for their use according to (ISO 14001, 1994).

4 WHAT CAN WE LEARN FROM THE QUALITY CIRCLE ?

Environmental problems can be handled only in a co-ordinated and systematic way by developing straight mechanisms. This has been realised lately by proactive companies.

The function overlapping integration gives a chance to recognise problems very early. In this way a minimisation of risks and a maximisation of chances is given.

Although the critical phases of a products live are very different within various industrial sectors, in a global overlook according to (McKinsey & Co., 1991) production and disposal/recycling have the strongest impact on the environment as Table 1 shows.

Table 1 Critical phases of product life by sector, in %

	Sourcing	Production	Product use	Disposal/recycling
Average:	11	47	13	29
Chemicals	10	49	21	20
Energy	13	58	16	13
Metals	8	62	3	27
Process industry	11	41	11	37
Durables	16	24	25	35
Consumer goods	7	45	9	39

The idea of a value chain is a systematical identification of ecological problems through all functions or parts of a company. The three basic functions - sourcing, production and distribution - include all company's functions. These three functions - added by all interfaces - can be very useful to build the basis for material balances as well as process and product balances for the problem identification. The value chain analysis is a diagnosis tool with its aim on cost reduction an yield increase.

The quality circle according to (EN ISO 9004-1, 1994) is a phase concept developed on all processes within a company (see Figure 1). As not only costs are in the point of view the interactions between the different stations are taken into consideration too. Therefore the result is a closed phase loop as Figure 1 shows.

Given by the models background of the involvement of all phases of a products life with its interfaces to the environment, the model is not only very universal it is also very useful for the analysis and optimisation of methods and processes in a single function or part of a company.

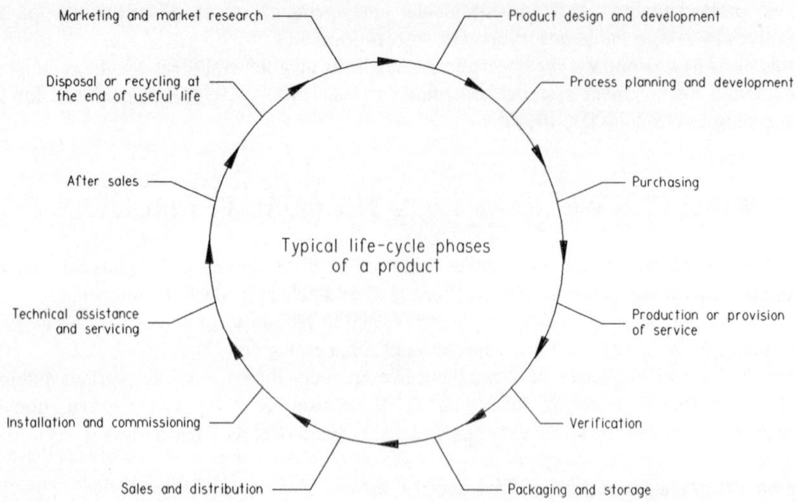

Figure 1 Quality circle.

5 WHAT SHOWS THE COMPARISON OF MANAGEMENT SYSTEMS ?

One of the most important actions on the way to an integrated environmental management system is the organizational embedding of environmental thoughts in a company's organisation based on an environmental strategy.

5.1 Philosphy, strategy and policy

The philosophy of a company includes the common aims and values within this organisation. Thus it is the starting point for the determination of the whole system.

Environmental ideas of companies are mostly formulated with reference to specific addresses for the society. With this the growing environmental awareness of the society was taken into account in the past. But facing growing environmental problems in the 1980's few companies widened their principles. Nowadays an exellent example for such principles is formulated in the ten guidelines of an already published environmental codex (International Chamber of Commerce, 1991). These can be easily adopted to a company-wide guideline for environmental care.

A strong similarity of these guidelines to the statements of quality management is evident. Both request an involvement of all relevant actions and tools within the policy, the possibility of public information and a system enclosing all hierarchical levels with the goal to provide continuous improvement. These guidelines leading to an overall principle for the company aim on an innovative environmental strategy.

Although there are a lot of different strategies discussed within publications they all can be reduced to three strategies. Figure 2 shows the matrix of these strategies which have different meanings within different industries and branches but they imply the same system.

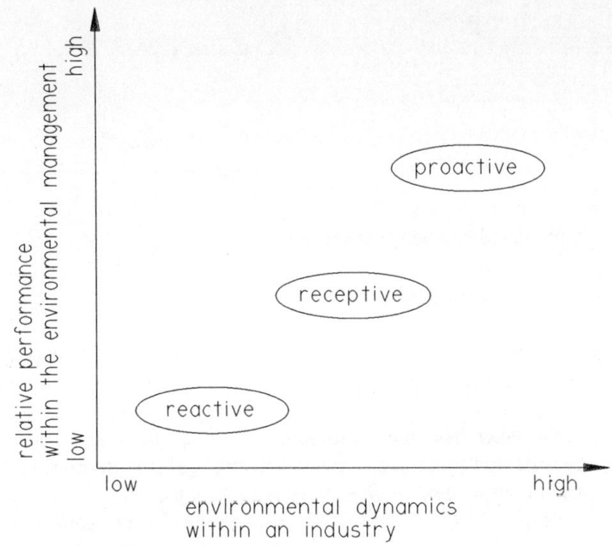

Figure 2 Matrix of environmental strategies.

5.2 Organizational structure

On the one hand as environmental care needs the participation of all employees the sequel is that there is no need for institutionalisation of environmental tasks because the implementation of such an environmental department takes away the personal responsibility for environmental tasks from the employees.

On the other hand the complexity of environmental decisions is a good reason for the centralisation of environmental planning and decision making on a higher hierarchical level. But this can lead to an overtax of this organisational level.

For the institutionalisation of environmental tasks can be given several reasons according to (Heiss, 1994):

- reduction of complexity
- balancing function
- control function
- innovation function.

A useful proposal for the embedding environmental functions shows Figure 3.

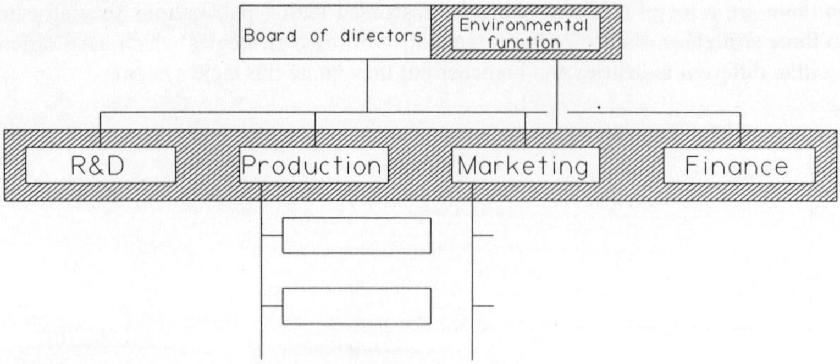

Figure 3 Proposal for the organisational company structure.

5.3 Quality circle and environmental circle

The idea behind "quality circles" in this context is that problems are most easily recognised and eliminated at their point of origin. This means that the possibility to solve problems within their working place is given to employees and supported by the membership of a problem-solving-group. As the name "quality circle" taken from the Japanese model of "quality control circles" is no longer state of the art - because nowadays this concept is in most cases no more used for quality relevant issues - we speak in this context of problem-solving-groups.

Environmental care within a company is only efficient if it is supported by everyone and all the small actions, which as a sum are a key-success-factor of an environmental quality management system, are done by each employee. Problem-solving-groups with specific environmental tasks are called "environmental circles".

- The implementation of environmental circles can bring a series of advantages:

Everyone is confronted and involved with environmental problems and therefore as part of the new philosophy and system involved persons are motivated.

- Enormous human capital is set free for environmental topics.
- A lot of small weak points are in sum a huge potential of failure and costs.

5.4 Audits

In the above mentioned publication (International Chamber of Commerce, 1991) a good definition of environmental auditing is given as a "management tool comprising a systematic, documented, periodic and objective evaluation of how well environmental organisation, management and equipment are performing with the aim of helping safeguard the environment".

This can be achieved by:

- facilitating management control of environmental practices and
- assessing compliance with the company's policies which include

meeting regulatory requirements.

There are similar definitions given in the new proposals for international standards (ISO 14000, 1994) and (ISO 14001, 1994). A strong correspondence to the definition for quality audits according to (EN ISO 9004-1, 1994) is obvious.

Additionally it must be taken into consideration that there exist also proposals for legal regulations for the auditing of environmental systems e.g. (Austrian Draft Law, 1994).

7 CAN WE DRAW ANY CONCLUSIONS ?

The advantages of an environmental management system based on a quality management system is not only the application of a widely accepted and approved system. Additionally, various efficient tools and methodes are ready for use.

On the basis of the guidelines for international quality management system a frame system for environmental issues can be easily established and implemented in a company. Many elements can be derived directly from quality management, other already established modules can be installed easily.

The joint characteristics of quality and environmental issues are the point of view of this paper. Different basic concepts and methods for subsolutions within the system are discussed and possibilities for implementation are proposed. Further the environmental management system with its system parts and the common use of some important quality tools in a modified form can build the basis for a system to handle the interactions between a company and its environment.

In this way it is possible to find an effective but low cost solution for an environmental management system especially under the point of view of small and medium sized enterprises. Recently published draft standards e.g. (BSI, 1994) show the possibility that the general ideas given in this paper may be of help for future tasks of organisations and enterprises in the field of health and security.

8 REFERENCES

Austrian Draft Law (1994) Bundesgesetz ueber die Zulassung von und die Aufsicht ueber Umweltgutachter sowie ueber die Fuehrung des Standortverzeichnisses entsprechend dem EU-Gemeinschaftssystem fuer das Umweltmanagement und die Umweltbetriebspruefung - Oeko-Audit-Gesetz. 144761/7-II/C/5/94.

BS 7750 (1992) Specification for Environmental Management Systems.

BSI (1994) Guide to Health and Safety Management Systems. Draft Standard, 94/408875DC.

EN ISO 8402 (1994) Quality Management and Quality Assurance - Vocabulary.

EN ISO 9000-1 (1994) Quality Management and Quality Assurance Standards - Part 1: Guidelines for Selection and Use.

EN ISO 9004-1 (1994) Quality Management and Quality Systems Elements; Guidelines.

Heiss, C.P. (1994) Managementloesungen fuer die Umwelt durch angepasstes Qualitaetsmanagement. e & i, **111**, 4, 189-192.

International Chamber of Commerce (1991) ICC Guide to Effective Environmental Auditing.

ISO 14000 (1994) Guide to Environmental Management Principles, Systems and Supporting Techniques. ISO/TC 207/SC 1 N 48, Committee Draft.

ISO 14001 (1994) Environmental Management Systems - Specification with Guidance for Use. ISO/TC 207/SC 1 N 47, Committee Draft.

McKinsey & Co. (1991) The Corporate Response to the Environmental Challenge. Summary Report.

Osanna, P.H. (1991) Qualitaetssicherung und Qualitaetsmanagement an der Universitaet. QZ - Qualitaet und Zuverlaessigkeit, **36**, 8, 459-464.

9 BIOGRAPHY

Prof. Dr. P.Herbert OSANNA is head of the Department for Interchangeable Manufacturing and Industrial Metrology (Austauschbau und Messtechnik) at the Institute of Production Engineerin, Vienna University of Technology (TUW, Karlspl.13/3113, A-1040 Wien, Austria). He is author of more than 150 scientific publications and he has given more than 150 lectures in Europe as well as in USA and Asia.

Dr. Daniel PROSTREDNIK studied at the Technical University of Bratislava. Since 1993 he is member of the staff of the Department for Interchangeable Manufacturing and Industrial Metrology.

Magister Angelika OSANNA, M.A., graduated in Physics at the University of Vienna in 1992. Since 1993 she is at the Department of Physics at the State University of New York at Stony Brook (SUNY, Stony Brook, NY 11790-3800, USA).

PART FIVE

Decision Support

15

A Teamwork Approach to Decision Support System Development for Watershed Management Problems

D.C.L. Lam[1], *D.A. Swayne*[2], *G.S. Bowen*[3] *and D.F. Kay*[1]

[1]*National Water Research Institute, Environment Canada, Burlington, Ontario, CANADA L7R 4A6, ph. 905-336-4916, fax 905-336-4582 email:David.Lam@cciw.ca and Doug.Kay@cciw.ca*
[2]*Dept. of Computing and Information Sci., U. of Guelph, Guelph, Ontario, CANADA N1G 2W1 email: dswayne@snowhite.cis.uoguelph.ca*
[3]*Watershed Management Section, Ontario Ministry of Environment and Energy, Toronto, Ontario, CANADA M4V 1L5 email:boweng@gov.on.ca*

Abstract

Decision-support systems (DSS) for solving complex environmental problems are not merely computer software systems. A working system requires data and knowledge inputs from domain scientists and the results need to be easily understandable to decision makers. It must be designed with both scientists and end users in mind, and must involve them early in the development, all working as a team. They need to overcome attitude and communication problems that exist among them and the system. This paper discusses the issues surrounding a team approach of building a DSS.

Keywords

Decision support system, knowledge-based system, environmental information system

1 INTRODUCTION

Environmental problems such as climate change and watershed management are multidisciplinary in nature. Solutions require the understanding of many disciplines of science: meteorology, hydrology, geochemistry and ecology, etc. With the advent of the computer, scientific data are now readily stored in computer databases and can be retrieved with sophisticated structured query languages (SQL), sometimes through fast computer network. The knowledge of the scientific processes, e.g. atmospheric transport or nutrient dynamics, are often encapsulated in computer simulation models.

As computer applications in data- and knowledge-bases increase, one would expect that the linkage between data and models is made more user-friendly. Significant efforts have been made (e.g. Loucks and da Costa, 1991) to promote the development of so-called decision-support systems (DSS) which can help, if not literally deliver the data and models to the hands of the decision-makers, at least demonstrate how the gap between science and policy could be brought closer. However, several obstacles remain in the development and operation of these decision support systems.

At the outset, the problem is not so much with the databases. It is always possible, given time, to convert from one data format to another. Yet, scientists and decision-makers still experience the so-called "information gridlock" syndrome. For as long as different agencies continue to collect environmental data, without coordination, it is difficult to find out how much data there are and who owns the data. Sometimes, the real hardship is to obtain the consent to use the data. Similar institutional problems exist with the use of models. It is often difficult for the designer of the decision-support system to predetermine which models to use. There are simply too many models to choose from, or scientists may offer their own favourite models instead of pre-selected ones. Thus, if the system is designed without these team considerations and the flexibility to accommodate "team" data and models, it will face strong resistance or "attitude" problems eventually.

2 THE RATIONALE AND THE APPROACH

One may challenge the notion that an open or teamwork-based DSS framework can actually change one's attitude. Our experiences over the past ten years show that DSS is basically a communication vehicle between scientists and policy makers, with the DSS developer in the middle. Typically one would start with the policy makers or the users to understand the problem to be solved and their expectation of how DSS may provide them with solutions. Then to bring in the data and models, one may need to talk to scientists familiar with the problem, since the DSS developer is usually not a domain expert. When relevant data and models are blended in with necessary supporting tools such as Geographical Information Systems (GIS), graphical interfaces and models, the prototype is to be tested by the user. More problems and requirements are then identified and the final product would be polished through several cycles of iterations with the scientists and the policy advisors.

Most scientists and policy advisors are cooperative since it is in their best interest to work toward a successful final product. However, since DSS development is about communication among people with diverse backgrounds, it is unavoidable that the communication may at times slow down because of lack of understanding and trust. For example, there was the tendency to hold back data, information or knowledge because the intent of the DSS was not explained clearly or scientists are afraid that their data and knowledge might be misused. Some believed that combining data and models from different disciplines would not generate new knowledge and therefore the DSS exercise was redundant. Some believed that they should be the one to be consulted directly by policy advisors, for fear that DSS could distort their knowledge if unsuccessful and could displace them if successful. Decision-makers or advisors, on the other hand, were generally concerned about whether there might be good, credible science implemented in the system. They were concerned about the cost, the socio-economical impact and the policy aspects of the decision. They were not interested in too much scientific details, but required a clear answer with limited quantifiers on uncertainties.

Given that the DSS is a team exercise, we have to deal with these human communication problems at the outset. Our experience is that the sooner we can build a prototype, the better we could use it to communicate among team members. Using examples, we can explain better the purpose and functionality of system collectively being built. We can then quickly overcome the fear and mistrust, if any, among the experts. We also provide the policy advisors a glimpse of what the DSS can do and let them provide an early input into the design of the system. The disadvantage is that the data and models may be too preliminary and the results too crude. However, we can turn this disadvantage into an advantage by showing the need for better data and models and challenging the experts to provide them. Indeed, we were able to obtain their data and models more effectively this way.

Thus, our approach is to develop a framework containing generic modules that are already tightly linked and allow information flow freely among the database, models, GIS, computer graphics and expert systems (Lam and Swayne, 1991). Recently, we have expanded the framework under the Windows system. By making use of advanced features in Windows, we are able to open more freely the connection to other databases and models. The user of such a framework can easily put their own data and models in the system and automatically linked up with those from other team members, without having to construct the linkages themselves. In this way, the team can devote more time with the problem than to struggling with information technologies.

3 THE FRAMEWORK

The RAISON system (Lam and Swayne, 1991) under the Windows environment is designed with this teamwork approach in mind. The system can be used to integrate environmental data for air, water, soil, sediments and biota or to couple models for these media so that scenarios for management options and cost optimization can be visualized with computer graphics and maps (Lam *et al.*, 1994). The system offers generic modules already linked to each other for manipulation of databases, models, graphics, GIS and optimization procedures, etc.

Figure 1 shows the framework for solving an environmental problem such as watershed management. Typically, information for each of the disciplines such as socio-economics, air issues, land, water, ecology and human health can be entered to the system and linked up through the database and GIS. For example, for watershed management problems, the industry may produce air pollutants or agricultural activities may lead to dispersion of pesticides, which may enter the terrestrial and aquatic regimes, thereby endangering fish species. The hydrological pathways can be modelled and the computed runoff can be used in chemical transport models to determine the ecological impact, etc. The ecological impact or damage cost can be used as the objective function in the overall optimization procedure (Fig. 1) that would select the best policy strategies (regulating water flows, expanding sewage treatment, farming practices, etc.) to achieve the preset goal. As well, the uncertainty estimation and error propagation for the connected models can also be determined to become part of the advice to the policy makers on how much confidence level that can be placed on the model predictions (Fig. 1).

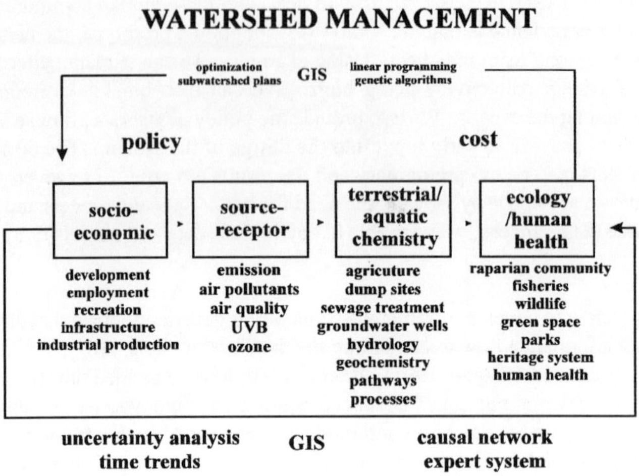

Figure 1 A framework for a decision support system for watershed management.

4 INTEGRATION OF DATA AND OBJECTS

4.1 Software design considerations

Data can be entered to the RAISON/Windows system through conversion interfaces available for many databases in the Windows system. GIS maps can be entered in vector and raster formats and can be used in overlay analysis of different map layers of information. For example, in Figure 2, information on fish can be retrieved from the database for a watershed and can be displayed against different background maps, e.g. stream and roads or geology (Fig. 2), from different sources. As expected, the trout species tend to thrive better in upstream locations because they prefer colder temperature, whereas a cursory, visual comparison of fish abundance data and geological factors (Fig.2) can lead to further study. It is this type of quick data visualization that helps promote new ideas and find new knowledge not easily attainable without the aid of such an integrated system.

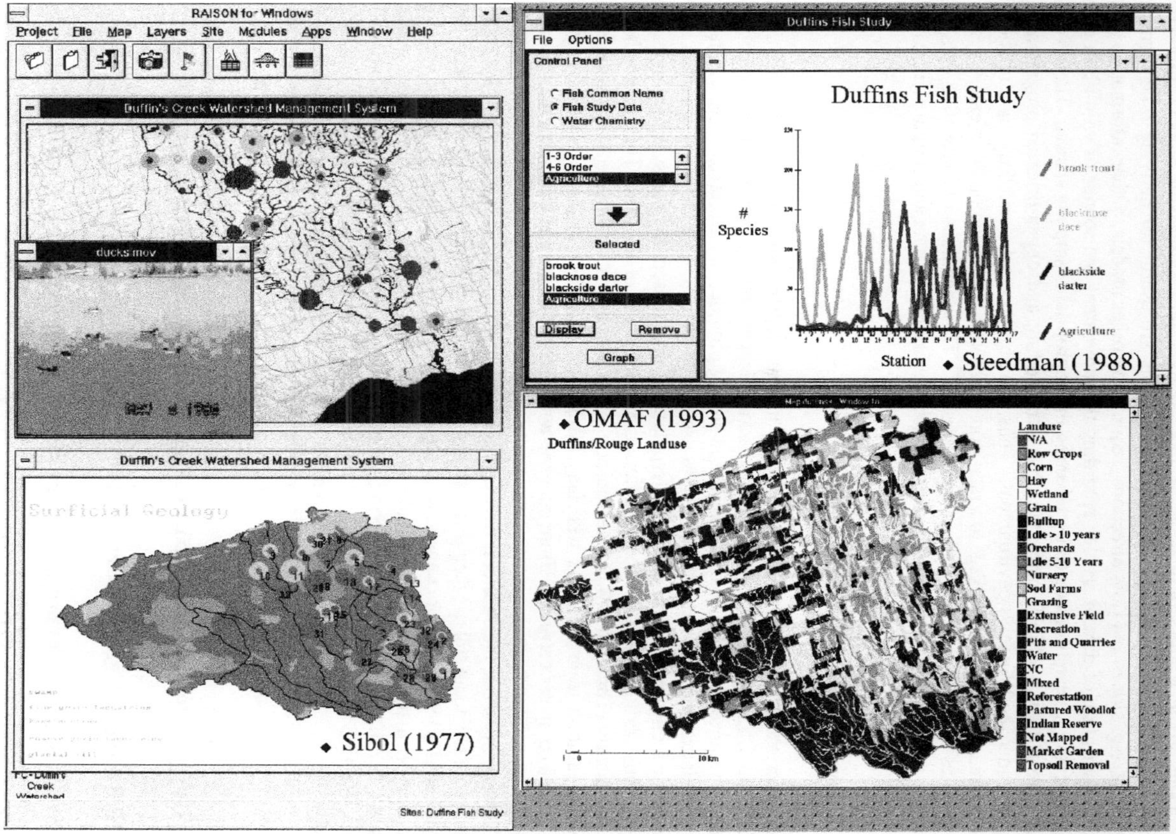

Figure 2 Data integration for the Duffins Creek Watershed, Ontario, clockwise from top left: Fish data overlaid on stream and road map with insert of a video of the wildlife habitat, fish data, land use map, geology map overlaid with fish data.

Similarly, maps on soil types and land use (Fig. 2) can be used to derive model coefficients for various types of hydrological models. In the RAISON system, special functions are developed and made available through a library of functions to automate algorithms for deriving model coefficients from GIS maps for these models. These GIS functions are useful for deriving other generic characteristics, e.g. slopes and flow directions, required by other models. Other information such as textual description of data or a video recording the movement and sounds of wildlife in their natural habitat (Fig. 2) can be accessed by selecting the appropriate sampling station. These video and items are stored as objects using the object-embedding and object-oriented programming techniques under the Windows environment. Thus, the objects can be called up via a relational database, by attached a dynamic linkage to a map icon, or through the execution of an expert system which activates selected objects based on model results. This technique of integrating data and objects serves well in bridging the gap between complex environmental data and simple pictures and images.

4.2 Team work considerations

From our experiences, it is important to show the identity of the data source once the data are made available in the DSS. Not only proper credits and acknowledgements are provided, but also contact addresses are required to facilitate communication among providers and receivers. An annotation window associated with each database is always useful to highlight the essential information such as the analytical chemistry method or the interpolation scheme used, the measurement units, the field conditions during data collection, and limitations in the use of the data. The system should provide the facility to bring in new data sets quickly, sometimes almost instantaneously during team workshops, among team members who may bring their own data in electronic forms or via local/wide area computer networks. From our experiences, the Windows environment allows many software products for database and spreadsheets to transfer data into a system such as RAISON easily. In many instances, the transfer is only temporary, via the Windows Clipboard into RAISON, so that team members can still use their own database or spreadsheet software to illustrate their data and use the RAISON system for connecting to other databases, maps or models. In this way, we provide a wider margin of comfort to those team members who prefer to use their own database systems and familiar routines, while offering the opportunity to integrate their data with others in RAISON. We find this approach quite effective in helping new team members overcome the initial institutional barrier. As they become more familiar with the team work approach and the RAISON system, they generally see the benefits of working together to solve practical problems, and then the data flow and accessibility are no longer temporary but more committed to the team.

5 INTEGRATION OF MODELS AND MANAGEMENT NEEDS

5.1 Software design considerations

The DSS framework provides an open structure for allowing different types of models to be linked with each other. For example, for nonpoint source pollution, a hydrological model is used to compute surface runoff which is then used as input to an agriculture nutrient/pesticide model. The nonpoint source inputs of nutrients and toxics are then linked to an advection-dispersion model that can also handle point source problems such as sewage effluent (Fig. 3). Further linkage to groundwater pollution models is possible. Under the DSS framework (Fig.1), these models can be incorporated into the system by: (a) introducing the source code as given or rewritten in a programming language compatible with C, C++ or Visual Basic and compiling it as part of the system executable, (b) using simple programming tools to develop an interface that intercepts the input and output data of the model into the DSS, but using only the executable code of the model (i.e. no need to rewrite the models), or (c) executing the model essentially on its own using an input file created *a priori*, saving the output in a database and then reading the output files from the DSS. The interaction of the model with the DSS is the greatest in (a) but the programming effort is correspondingly the highest. The opposite is (c) where both the interaction with DSS and the programming effort are the least. From our experience, since method (b) offers the compromise between DSS interaction and programming effort, we used it in most cases. Alternatively, if the model inputs and outputs themselves are sufficient to support the creation of a simple input-output model, we can use the input-output model so derived to emulate the original model and, if possible, to use it in the DSS for the specific application. The results may be approximations but the DSS interaction would be the fastest, even faster than the original model.

The need to have faster interaction with other models in DSS is obvious when management strategies are considered. For example, when designing the flowpath and reservoir network, or when establishing the cost of expanding an existing sewage treatment plant to achieve a downstream water quality standard, the models will be run hundreds of times as they are nested in an optimization procedure (Fig. 3). For engineering design problems, such rigorous investigation is required. For some management consideration, however, particularly when the objectives, e.g. ecological impact, are fuzzy and the cost estimates are crude, less rigorous answers may be acceptable. In the latter case, the use of artificial intelligence techniques such as genetic algorithm, fuzzy expert system and causal network (Lam and Swayne, 1994) can offer reasonable iteration rates and are flexible for generating practical answers without greatly compromising the science. In the end, what the scientific and socio-economic team can deliver to the decision-makers is a system composing of their joint data- and knowledge-bases with only a few menu buttons for displaying simple results and recommendations, including the advice on decision and approval relating to possible actions, e.g. sewage treatment plant expansion (Fig. 3).

5.2 Team work considerations

For the benefit of the team, it is important to identify the source of the models and to provide an annotation of their assumptions, limitations and applicability, once they are incorporated as part of the DSS. Generally, modellers tend to promote their own favourite models, either by virtue of being the authors themselves or simply being more familiar through practice. The institutional barrier is therefore great when one modeller claims to have a better model than others, often arguing over whose model has a better process formulation, a finer scale resolution, or a more accurate computational method. However, while these are basic elements for constructing a model, the choice of the most appropriate model for solving a practical problem often depends on the questions being asked (Lam and Swayne, 1991). For as long as the model can answer the questions within

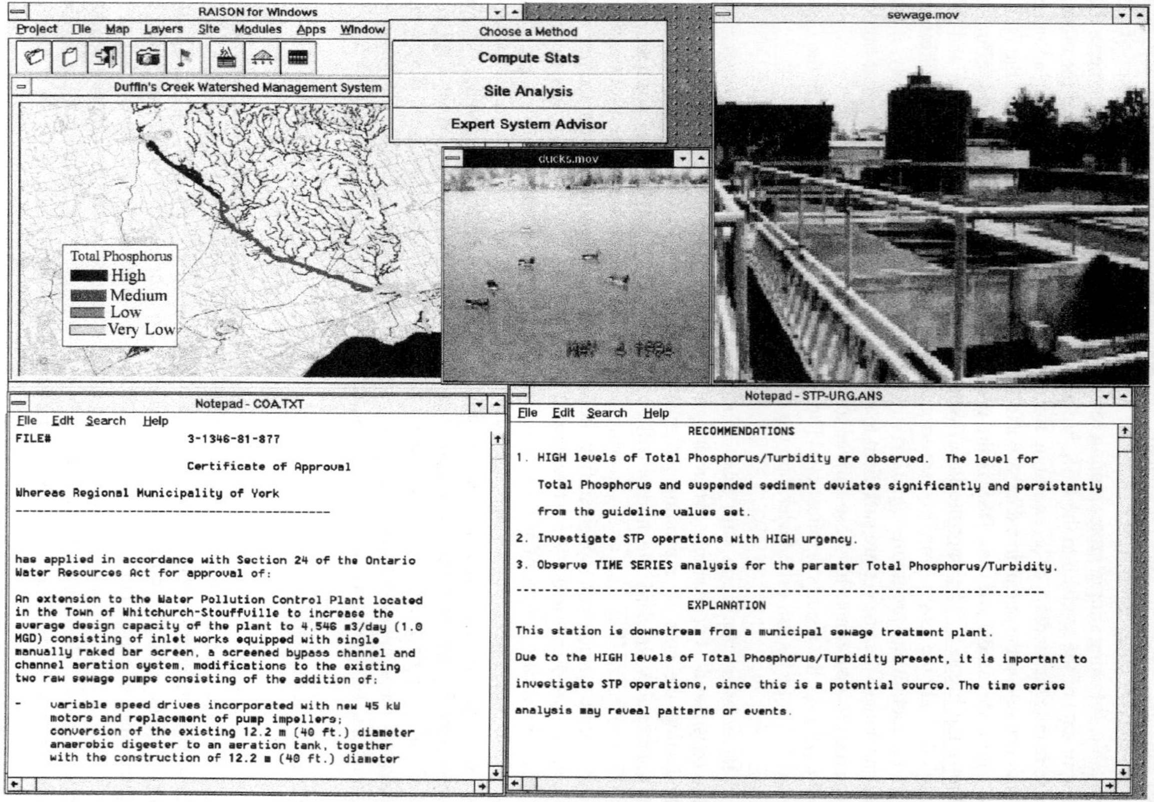

Figure 3 Models and expert advice for sewage treatment effluent in Duffins Creek Watershed, Ontario, clockwise from top left: Pollutant transport results, simple menu to direct result display, sewage treatment plant in question, recommendations based on data and model results, an excerpt of certificate of approval.

a reasonable tolerance on the accuracy of assumptions, scales and methods, it is a good model. A good model can be applicable in an integrated assessment only when it fits within the tolerance of, and is capable to link to, models selected likewise from other disciplines. In other words, not only do the modellers have to learn models developed in their own and other disciplines, but also they have to accept the integrated assessment perspective in determining which model is more appropriate. To evercome this barrier, our experiences show that, by and large, when the assumptions, scales and methodologies of the models, regardless of their original disciplines, are presented, a short list of eligible models emerges. To select the most appropriate among these eligible models, an expert system is used, first by going through the assumptions and applicability of the models and then building a rule base for selecting models based on available data and knowledge, with the full participation of the modellers and other experts in the team. Thus, the rule base is transparent and understandable to everyone, and the models are chosen based on team advice and available information. Similar team considerations are needed in selecting constraints and criteria for the optimization of resources and scenario testing for management strategies, although the rules are probably more uncertain and fuzzier than the case of model selection.

6 CONCLUSION

Over the past ten years, we have developed a generic DSS framework to help solve a wide range of environmental problems, from acid rain and mine effluent to watershed management and climate change. What impressed us most during these ten years are not so much the rapid advances in information technologies. It is the day-to-day attitude problem among team members and the communication problem between human and the machine. If these team building problems are ignored, the DSS remains an empty shell. Future information technology development should therefore be directed to facilitate this team building exercise: the assessibility of databases and maps through fast networks, the better documentation and linkage among models and the "soft" engineering tools that allows qualitative knowledge to be formally and critically used alongside with models and other decision support methodologies.

Acknowledgements. The authors thank E.D. Ongley and P. Miller for their support and the research partners in this project for generously providing data and knowledge.

7 REFERENCES

Lam, D.C.L. and Swayne, D.A. (1991) Integrating database, spreadsheet, graphics, GIS, statistics, simulation models and expert systems: experiences with the RAISON system on microcomputers, in *Decision Support Systems - Water Resources Planning* (ed. Loucks and da Costa), Springer-Verlag publ., Berlin, p. 429-460.

Lam, D.C.L. and Swayne, D.A. (1994) A hybrid expert system and neural network approach for environmental applications. *Proc. 2nd Int. Conf. Expert Systems for Development*, (Ed. R. Sadananda), IEEE Computer Society Press, Los Alamitos, California, p. 298-303.

Lam, D.C.L., Mayfield, C.I., Swayne, D.A. and Hopkins, K. 1994. A prototype information system for watershed management and planning, J. Biological Systems, **2** (4), 499-517.

Loucks, D.P. and da Costa, J.R. 1991. *Decision Support Systems - Water Resources Planning*, Springer-Verlag Publ., Berlin, 574 pp.

8 BIOGRAPHY

David C.L. Lam received a Ph. D. in Computer Science from the University of Waterloo in 1974. He is currently a Senior Scientist and Project Chief, Environmental Information Technology, New Technologies Research Branch at the National Water Research Institute of Environment Canada. He was also a part-time Professor in the Civil Engineering Department, MacMaster University, and is currently an Adjunct Professor in the Computer Science Department, University of Waterloo, Canada. His research interests are in environmental modelling, "soft" engineering, decision support system and environmental information system.

David A. Swayne received a Ph.D. in Computer Science from the University of Waterloo in 1975. He is a member of the Society for Industrial and Applied Mathematics. He is currently a Professor in the Computing and Information Science Department and the School of Engineering at the University of Guelph, Canada. He is also an Adjunct Professor in the Computer Science Department at the University of Waterloo. Originally a specialist in numerical software, his research has focussed since 1983 on applications of computing science and information technology to the environment.

Gary S. Bowen has a B. Sc. degree from Trent University, Canada, and is currently a watershed management specialist with the Watershed management Section, Science and Technology Branch, Ontario Ministry of the Environment and Energy.

Douglas F. Kay has a B. Sc. degree from University of Guelph and is currently a Physcial Scientist with the New Technologies Research Branch, National Water Research Institute, Environment Canada. He specializes in data visualization and ecological systems.

16

Partially Ordered Sets - A Computerized Tool to Compare Environmental Databases

Brüggemann, R., Voigt, K.
GSF - Forschungszentrum für Umwelt- und Gesundheit
Projektgruppe Umweltgefährdungspotentiale von Chemikalien (PUC),
Neuherberg, 85758 Oberschleißheim, Germany
Telephone: ++49-89-3187-4030, Fax.: ++49-89-3187-3369
e-mail: VOIGT@wiz.uni-kassel.de

Halfon, E.
National Water Research Institute
Canada Centre for Inland Waters
Burlington, Ontario, Canada L7R 4A6
Telephone: ++1-905-336-4917, Fax.: ++1-905-336-4972
e-mail: u044@cs.cciw.ca

Abstract

A set of objects can be analyzed by several tools. When ranking is a concern, then lattice theory and its graphical representation (Hasse diagrams) are useful. This paper introduces a new approach to analyze Hasse diagrams with respect to ranking in the environmental field. Simple set theoretical and lattice theoretical concepts such as successor sets, their cardinalities and the intersection of these sets have been used to compute a new matrix **D** that identifies the relation among objects. As an example we analyze a published data set of 22 objects (online databases of chemical and environmental relevance). These objects are scored using a six number scoring system with five environmental and chemical attributes. The online databases Toxline, Environmental Research Projects and Chemical Compliance Monitor turn out to be the "best" that is to say maximal elements in this scoring approach and within the concepts of lattice theory. Furthermore the importance which attribute contributes for the ranking procedure is shown. For this purpose we have introduced another new matrix, **W**, which quantifies the dissimilarity of different Hasse diagrams and a sensitivity measure s(*E*) that analyzes the importance of criteria, by which the environmental databases are characterized. A crude estimation of the upper bound of s(E) is given. It can be shown thatthe attributes: "information parameter for environmental chemicals" (IP) and "number of chemicals" (NU) are relatively important and the attribute "Use of chemicals" (US) is relatively unimportant for the ranking of the chosen 22 environmental databases.

Keywords
Environmental databases, Hasse diagrams, ranking, sensitivity analysis.

1 INTRODUCTION

The basis of the Hasse diagram technique is the assumption that a ranking can be performed while avoiding the use of an ordering index (Halfon, 1986). In our application, Hasse diagrams present information not only on ranking but, most of all, they show whether the criteria, characterizing the objects, lead to ambiguities in the ranking procedure: For example, an object might be ranked higher according to one criterion but lower according to another. These two objects are not ordered because their data are "contradictory" to each other. This ambiguity -which is important to know for further applications- is not evident when we use an index for ranking, but it is immediately evident in a Hasse diagram. Hasse diagrams are extremely useful if several criteria are given to decide which objects are priority objects. In our approach the objects are environmental databases. The criteria are environmental and chemical ones and will be given later. The scoring is carried out according to a six number scoring system. Here we can not only identify the priority databases but furthermore it can be shown that these databases have different qualities.

In this paper we investigate a method to extract information from Hasse diagrams since the casual user of the ranking method of Halfon (1986) might become confused by the complexity given of the graphical information in a Hasse diagram and might not be able to use it. Furthermore, we study the influence of the choice of criteria to rank a set of objects (precise definitions follow later). **The ranking of a set of objects does not only depend on the numerical values, but even more on the choice of criteria.**

The results of this analysis are two matrices, **D** and **W**, that identify the main features of the structure of Hasse diagrams and quantify the influence of criteria on the ranking.

2 HASSE DIAGRAM TECHNIQUE APPLIED IN ENVIRONMENTAL PROTECTION

The Hasse diagram technique has been already applied in several fields to rank objects of environmental relevance (see for example Brüggemann, 1995 b, c and Voigt, 1994).

3 THEORY OF HASSE DIAGRAM TECHNIQUE

Davey and Priestley (1990) present useful background information on graphs, sets, partially ordered sets (**posets**) and Hasse diagrams. A few theoretical explanations will be given in the following paragraphs for a better understanding of the ranking procedure.

3.1 Hasse diagram

Hasse diagrams visualize the order relations within posets. Two objects x, y of a poset are ordered if all scores of x are less or equal than those of y. Hasse diagrams are oriented graphs (digraphs). A digraph consists of a set *E* of objects drawn as small circles in Hasse diagrams. In our applications the circles near the top of the page (of the Hasse diagram) indicate objects that are the "best" objects according to the criteria used to rank them: These objects have no predecessors (they are not "covered" by other objects) and are called maximal elements, abbreviated also to "maximals."

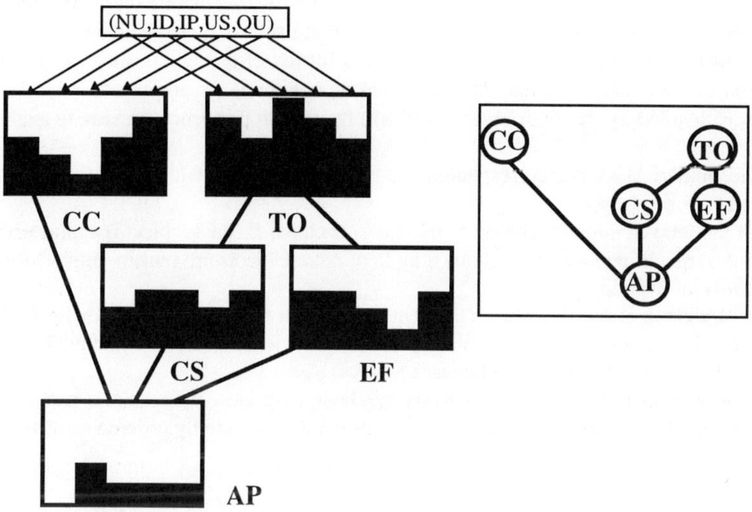

Figure 1 Two representations of a simplified Hasse diagram.

A line in the Hasse diagram indicates that the two objects connected by that line are "comparable" with each other, lack of sequences of connecting lines indicates that there are contradictions (a complete explanation with examples may be found in Halfon, 1986). A simple Hasse diagram will exemplify the above given statements; it is a part of the later discussed full diagram.

In Figure 1 the left diagram is an extended form of a Hasse diagram, where the scores are shown in the form of bar diagrams. We use the same scores 0 to 5 as in the full diagram (see below). The assignment of the attributes NU, ID, IP, US, QU (see chapter 5.1) to the bars is indicated by the arrows in the upper part of the left side diagram. The objects (environmental databases) are

CC = Chemical Compliance Monitor
TO = Toxicology Information Online
CS = Chemical Safety Newsbase
EF = Environmental Fate Database
AP = Air Pollution Technical Information Center

For example the two databases AP and TO are ordered, i.e. AP ≤ TO.
On the right side the usual shorthand representation of Hasse diagrams is shown, the assignments of the scores to the objects are usually given in a table.

3.2 Definitions of basic terms

Criteria include both quantitative and qualitative properties.
An **attribute** is a quantitative, measurable, criterion. Generally, we denote these attributes as #1, #2, ..., #n. It is convenient to denote the full attribute set as *A*. We call the full set of attributes an **information basis** (IB) of the ranking. A **family** of $p := 2^n - 1$ attribute sets will be considered in our analysis, namely the **power set** of *A* without the empty set. Each subset of attributes is denoted by A_i , with $A_i \subset A$ and will be used to perform a sensitivity analysis (see later).
Data are the numerical values corresponding to each criterion by which a given object is characterized.
An **object** is the item of interest. The set of m objects is called *E*. Each object is characterized by pairs, triples, quadrupels or generally by a tuple of data. Objects are ranked graphically by Hasse diagrams, as explained in 3.1.
Equivalent objects in Hasse diagrams: Different objects that have the same data with respect to a given set of attributes. They are elements of an equivalence class with the equivalence relation "equality of the characterizing tuples."
A **case** is a shorthand notation for an analysis by Hasse diagrams of m objects and with a defined attribute set A_i. Thus, a given set of attributes induces a partially ordered set, which is visualized by a Hasse diagram.

3.3 Definition of the terms "key elements" and "succcessor sets"

Substructures within a Hasse diagram, i.e., relations among objects as well as the importance of criteria in ranking are investigated with the help of key elements and successor sets. Any object of the poset can be chosen as a starting point to begin the analysis. We call this object, a "key element". For convenience, all chosen key elements are considered to be elements of a set *K*, a subset of *E*.
In a Hasse diagram objects are connected by lines. Analysis of the successors of a key element implies a search of **all** objects located lower than the key **element and connected to it by a path, being a sequence of connecting lines** or equivalent to that of the key element. The set of all successors of a key element k is denoted as *G*(k). The properties of the

successor set $G(k)$ and its relation with successor sets of other key elements are first used to analyze the structure of a Hasse diagram, later they will be used to perform the sensitivity analysis. We write the cardinality of the successor set $G(k)$ as card $G(k)$. The cardinality of a set is its number of objects.

3.4 Relations between elements and posets

In order to investigate the global structure of the relations between any two elements of the posets, we perform this analysis mathematically. This structural relation might be hidden within the geometrical representation of the Hasse diagram. Thus, we introduce the symmetrical matrix **D**, the entries of which are calculated from the cardinalities of all intersections of pairs of successor sets:

$$D_{ij} := \text{card}\,[G(i) \cap G(j)]. \tag{1}$$

The matrix **D** may be the basis to define several quantities describing the relation between objects within the posets quantitatively and is related to the sensivity analysis. Here we use entries of the matrix **D** to look for:

- the representativity of any key element "i" which is - if exactly one poset is considered - given by D_{ii} and
- the number of common elements between any two key elements i and j.

Both numbers give the wanted rough insight into the Hasse diagram. We call this search for relations between objects an "unfolding" of the structure of the Hasse diagram.
Especially D_{ij} $(i \neq j)$ is of interest:
If D_{ij} has a high value then the two key elements have many successor elements in common: they are comparable to many objects with respect to all attributes.
If D_{ij} has a small value then the two key elements may be taken as representatives for two quite different sets of objects.

4 HASSE PROGRAM

The theory of the Hasse diagram technique is transfered to a DOS-Program Hasse, written in C. A WINDOWS-Version is anticipated.
The program is able to perform the following facilities:

- file organisation
- text-editing-functions
- data manipulation
- general characteristics of the posets
- visualization of posets by the corresponding Hasse diagram
- calculation of matrix **D** which entries may be used to study relations between key elements. We avoid the constructions of explicitly measures of similarity because further studies should be performed to establish an appropriate definition.
- setting-up and manipulation of diagrams
- performing of sensitivity analysis (standard and expert mode).

The program is driven using pull-down-menus and comprises a help function which could still be more extensive. The Hasse program is in the test period at the moment but will be put on the market in the near future (Brüggemann 1995a).

5 RANKING OF ENVIRONMENTAL DATABASES APPLYING HASSE DIAGRAM TECHNIQUE

5.1 Ranking of 22 Online Databases

Since 1993 we develop priority setting concepts for data-sources, using the scientific background of lattice theory and its evaluation by Hasse diagrams.
Voigt and Brüggemann have scored 22 environmental online databases using the following five environmental and chemical-relevant evaluation criteria (Voigt, 1994):

- number of chemicals (NU)
- identification parameters (ID)
- information parameters for environmental chemicals (IP)
- use of chemical substances (US)
- quality of database (QU) (Voigt, 1994).

The scoring is carried out according to a six number scoring system; a high value means a high quality whereas a low value has the meaning of a low quality with respect to the criterion used.

In Figure 2 the Hasse diagram of the poset of 22 online databases evaluated by the information basis { NU, ID, IP, US, QU } is shown. The interpretation of the diagram might be difficult because of the many circles and lines. The concept of matrix **D** to this Hasse diagram is applied not only to explain its use but also to facilitate the interpretation of Figure 2. In this diagram three **maximals** exist. These are the databases UF (Umweltforschungs-Datenbank, Environmental Projects Database), CC (Chemical Compliance Monitor) and TO (Toxicology Information Online). These databases (objects) are now defined to be the **key elements**. In Table 1 we list the elements of the successor sets.

Table 1 Successor sets of the databases TO, CC and UF

TO	*CC*	*UF*
AP, AQ, AR, AS, BA CB, CS, EB, EF, EN EV, GE, NE, OC, PO SC, UL, WA, WR	AP AR BA EN NE	AR BA EB EN SC
19	5	5

The cardinalities of the three successor sets are:
card G (TO) = 19
card G (CC) = 5
card G (UF) = 5

5.1.1 Hasse Diagram for 22 Online Databases

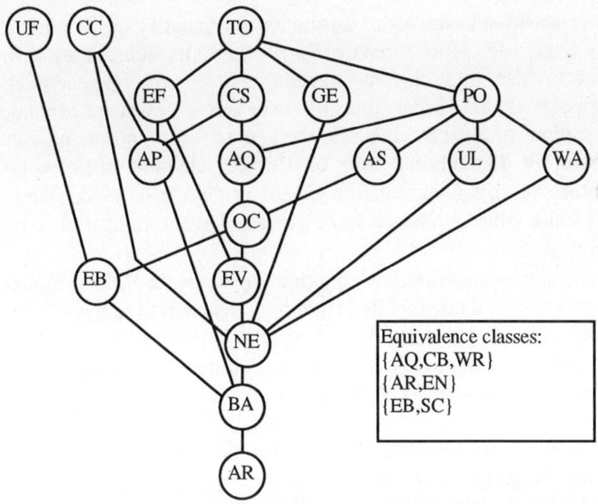

Figure 2 Hasse diagram for 22 Online Environmental Databases.

5.1.2 Example to use the entries of matrix D

The part of the intersection matrix **D**, referring to the three objects (i.e. K = {TO, CC, UF}) of interest is:

			1	2	3
			TO	CC	UF
	1	TO	19	5	5
D =	2	CC	5	5	3
	3	UF	5	3	5

The difference of the cardinalities of the two successor sets of TO and CC is 14 ($D_{11} - D_{22}$). TO has more successors (19) than CC (5). Therefore we conclude that even if TO and CC are ranked as maximals, TO represents more other databases which can be compared with this maximal than CC which has only 5 successors.

The cardinality of the **intersection** ($D_{12} = D_{21}$) of these two successor sets is 5, the same as the cardinality (D_{22}) of *G* (CC) which is 5. We conclude that even if the attributes of TO and CC make these two databases incomparable to one another, those databases represented by CC, namely AP, NE, BA, AR, EN are also comparable to that of TO. Similar argumentation can be found for TO and UF.

5.2 Sensivity Analysis of Ranking with Respect to its Attributes

5.2.1 Theory of Sensitivity Analysis

The ranking of the objects is sensitive to the set of attributes. To quantify the importance of an attribute on ranking the basic idea is to compare Hasse diagrams induced by different attribute sets with each other. In order to do this, the concept of successor sets generated by a key element is the starting point again. The notation of successor set must be expanded to include all the actual combinations of attributes. Within the generalization of having a family of attribute sets **the successor set depends not only on the key element but also on the choice of attributes.** Therefore the following notation $G(k,A)$ or $G(k,A_i)$ is used, where G is the successor set, k denotes some arbitrary chosen key element, A is the full set of attributes and A_i is a subset of attributes, $A_i \subset A$.

The influence of each attribute can be quantified by counting the elements of the symmetrized difference between two successor sets, their so-called Hamming distance as follows:

$$W(k,A_i,A_j) := \text{card}\{ [G(k,A_i) \setminus G(k,A_j)] \cup [G(k,A_j) \setminus G(k,A_i)] \} \quad (2)$$

For a given key-element, two Hasse diagrams (given by two arbitrary sets of attributes) are more dissimilar, the more the successor sets $G(k,A_i)$ and $G(k,A_j)$ differ.

To simplify notation, we write W(k,i,j) for $W(k,A_i,A_j)$. We also note that each

$$W(k,i,j) \geq 0 \text{ and } W(k,i,j) = W(k,j,i). \quad (3)$$

The matrix itself is denoted by **W**(k). Thus we have:

$$\mathbf{W}(k) = \begin{matrix} W(k,1,1), W(k,1,2), ..., W(k,1,p) \\ W(k,2,1), W(k,2,2), ..., W(k,2,p) \\ .... \\ W(k,p,1), W(k,p,2), ..., W(k,p,p) \end{matrix} \quad (4)$$

with $p = 2^n - 1$ (see definition of attribute). This matrix **W** is the key for the sensitivity analysis of ranking, each entry of **W** is the cardinality of the symmetrized difference (Equation 2) of two successor sets which are constructed from a given key element k and the two Hasse diagrams induced by two attribute subsets. This large but symmetrical matrix does not need to be analyzed in its entirety all the time because we are only interested in some few attribute sets (the standard mode of sensitivity analysis which is explained here). The sensitivity analysis can be performed by the following steps:

1) As we are only interested in comparisons of the full attribute set A with subsets A_i, only one row of the matrix **W** is of interest. We can choose the first one without loss of generalization, thus we are left with W(k,1,1), W(k,1,2), ..., W(k,1,p), where the index 1 denotes the full attribute set A.
2) To see the influence of attributes on a Hasse diagram we compare the Hasse diagrams induced by A with those induced by the attribute sets with only n-1 attributes. Therefore the effect of dropping exactly one attribute on the ranking is given by the remaining n entries of the first row, W(k,1,2), ..., W(k,1,n+1).
3) The remaining n matrix elements of the first row $W(k,A,A_1)$, ..., $W(k,A,A_n)$ are put together to form a "sensitivity tuple" of the key element k:
$\sigma(k) := (s(k,1), s(k,2), ..., s(k,n)) = [W(k,A,A_1), ... W(k,A,A_n)]$.
Note that the enumerations of the subset A_i are thought of as follows:

$A_1 = \{\#2, ..., \#n\}$
$A_i = \{\#1, ..., \#i\text{-}1, \#i\text{+}1, ..., \#n\}$ (5)
$A_n = \{\#1, ..., \#n\text{-}1\}$

4) The larger the entries s(k,i) the larger is the symmetrized difference between $G(k,A)$ and $G(k,A_i)$ and correspondingly the larger the influence of attribute #i on the position of key element k within the Hasse diagram under A compared with that under A_i.

5) The matrix W(k) depends on the selection of the key element k. If however, more objects are to be analyzed we generalize as follows:

$$W(K,i,j) = \sum_{k \in K \subseteq E} W(k,i,j) \quad (6)$$

where K is any set of key elements, in a shorter notation $\mathbf{W}(K) = \sum \mathbf{W}(k)$.

6) <u>All</u> objects are selected as key elements. Therefore instead of W(k), W(E) is to be investigated. W(E) is the total matrix of the set E. We note that a crude upper limit of W(E,i,j) can be found simply by comparing a poset of m solely non-comparable elements with a poset where all m elements are equivalent to each other:

$$0 \leq W(E,i,j) \leq m\,(m-1) \quad (7)$$

7) **W**(E) will be used as a measure of sensitivity. Accordingly we suggest to quantify the sensitivity by:

$$s(E,i) = W(E,A,A_i) \quad 1 \leq i \leq n \quad (8)$$

with the enumeration scheme of (5). According to (7), s(E,i) has values between 0 and m (m-1).

The tedious calculation of W (E, A, A_i) by the defining equation (2) is not necessary. It can be shown that the following equation is valid (Brüggemann and Halfon: publication in preparation).

$$s(E,i) = \sum_j (D_{jj}(A_i) - D_{jj}(A)) \quad (9)$$

where D (A_i) and D (A), denotes the matrices **D** under attribute A_i and A.

5.2.2 Application on the Example Hasse Diagram

The entries of matrix **W**(E) corresponding to the comparisons of the successor sets induced by the full attribute sets and of those induced by n other attribute sets where one attribute is omitted, are shown in Table 2. If we compare all attibutes used here, the attribute US (Use of chemicals) has a low, whereas the attributes IP (information parameter for environmental chemicals) and NU (number of chemicals) have a high influence on the ranking by Hasse diagrams.

Table 2 Results of the sensitivity analysis (standard mode)

full attribute set compared with case No.	attribute sets	omitted attribute	s(E,i)
2	A, A_1	QU	20
3	A, A_2	US	11
4	A, A_3	IP	34
5	A, A_4	ID	16
6	A, A_5	NU	34

Such finding is of importance for the practical use of the ranking procedure. In our example we are looking for priority environmental databases. Their identification depends on the actual criteria used. If NU is omitted, the analysis tell us that there are more changes in the diagram then by leaving out the criterion US. If IP is left out, the whole diagram changes, however the number of priority databases remains. Beyond this, it is of interest to discuss the attributes themselves and their role on the ranking procedure. A less important attribute may result from a positive correlation with another attribute.

6 FURTHER DEVELOPMENTS

The concept of partially ordered sets and their visualization by Hasse diagrams turns out to be very useful in many applications, where evaluative considerations are important. This applies to a large extent to environmental protection subjects. A recent - but brief - overview may be found in Brüggemann 1995 b, c. Considerable work is done to use the concept of posets as a modeling tool (Brüggemann, 1995 b) and to derive from posets elements of artifical elements (Wille 1987). At least the underlying mathematical theory of lattices allows to deduce a system of implications (rules). However the applications of this concept for environmental purposes is on its very beginning, but is going to be intensified by our working groups in Germany and Canada in the near future.

7 REFERENCES

Brüggemann, R., Halfon, E. (1995a) Theoretical Base of the Program Hasse, GSF - Forschungszentrum für Umwelt und Gesundheit, Neuherberg, in print.

Brüggemann R., Zelles L., Bai Q.Y., Hartmann A. (1995b) Use of Hasse Techniques for Evaluation of Phospholipid Fatty Acids Distributtion as Biomarkers in Selected Soils, *Chemosphere*, accepted.

Brüggemann R., Schwaiger J., Negele R.D. (1995c) Applying Hasse Diagram Technique for the Evaluation of Toxicological Fish Tests, *Chemosphere*, in press.

Davey, B. A. and Priestley, H.A. (1990) Introduction to Lattices and Order. *Cambridge University Press*, Cambridge.

Halfon, E. and Reggiani, M.G., (1986) On ranking chemicals for environmental hazard. *Environ. Science Tech.*, **23**, 1173-1179.

Voigt K., Brüggemann R. (1994) Umwelt-relevante Online Datenbanken im Vergleich, in *Gesellschaft Deutscher Chemiker, Fachgruppe Chemie-Information-Computer (CIC), Mitteilungsblatt* Nr. 29, GDCH, Frankfurt am Main, Mai 1994. 33-57.

Wille R. (1987) Bedeutung von Begriffsverbänden, in Ganter B., Wille R., Wolf K.E., *Beiträge zur Begriffsanalyse* 161-211, BI Wissenschaftsverlag, Mannheim.

8 BIOGRAPHY

Brüggemann, Rainer

Rainer Brüggemann studied chemistry in Munich and made his Ph.D. 1977 in Theoretical Chemistry at the University of Munich. From 1980 to 1983 he worked at the Technical University of Berlin on the subject of safety studies for nuclear waste disposal sites. In 1984, he moved to the GSF National Research Center for Environment and Health in Neuherberg. There he worked on risk analysis and ecological modelling and QSAR in the field of chemicals. In 1992 he became head of the project group on environmental hazard assessment for chemicals. He published more than 100 articles, reports and books on chemistry and several environmental issues.

Voigt, Kristina

Kristina Voigt graduated in food chemistry at the Technical University in Berlin in 1979. Then she worked at the German Environmental Protection Agency (UBA) in Berlin. In 1983 she moved to the GSF National Research Center for Environment and Health in Munich. There her main task is testing and evaluation chemical-relevant and environmental-relevant data-sources. Since 1988 she has been head of the research project "Information System for Environmental Chemicals". She has published extensively in books, reports, proceedings and journals. She holds training courses, seminars and lectures. She is a member of the editorial board of Online & CD-ROM Review.

PART SIX

Distributed Environmental Information

17

Environmental Information for Public Access

Peter Kutschera, Gerald Schimak, Heinrich Humer
All Austrian Research Centre Seibersdorf
Dept. of Information Technology
Kutschera@zdfzs.arcs.ac.at, Schimak@zdfzs.arcs.ac.at,
Humer@zdfzs.arcs.ac.at

Abstract

After the installation of the Austrian Ozone Monitoring Network in April 1994, it was soon clear that the information gathered all over the country should be provided to all interested in that matter. For that task an acceptable technology was found by using World Wide Web. To demonstrate how this new technology can be used to access actual environmental information via meta information the Austrian Research Centre Seibersdorf developed an WWW prototype.

In the following chapters a World Wide Web (WWW) gateway for the Austrian Ozone Monitoring Network will be described, first emphasizing some aspects of the Austrian Ozone Monitoring Network and second talking in detail about the implemented data presentation for environmental information.

Keywords

distributed environmental information system, air pollution monitoring system, visualization, integration, internet

1 SNAPSHOT OF THE OZONE MONITORING SITUATION IN AUSTRIA

At the moment there are about 180 measuring stations feeding continuously meteorological and pollutant information into the Austrian Ozone Monitoring Network. This stations integrated in local air quality measurement networks are operated by the responsible authorities in the nine Austrian federal countries. An additional tenth is operated by the Federal Environmental Agency Austria (FEAA) as nationwide background monitoring system.

To overcome the differences in the air quality measurement networks, the complexity of distributed data as well as heterogenity within the network the Austrian Research Centre Seibersdorf developed a solution to harmonize access to all of this stations within the network. The keywords for the new technology used are ozone network computer, meta information system, meta data distribution and integration of different environmental information. These methods are exhausted discussed in (Schimak 1994a and Schimak 1994b). Whenever there is a new or modified data in an air quality measurement network's central computer system the data is automatically transformed into the ozone data network format and sent over a local area network to the ozone network computer, where the values are stored in a local database. Thereafter the information is accessible by all remote ozone network computers. The measured, aggregated and stored values will be accessed by a high speed retrieval mechanism (e.g. formula server) installed on each network node. All ozone network computers are linked via public X.25 packet switched data network. Technical information about the server concept you will find in (Schimak 1994a and Schimak 1994b).

The continuously available information within the Austrian Ozone Network can be scaled in meta information and average values, like half hour mean values.

1.1 Meta information

Meta information can be described as information about information. It is a special kind of descriptive data and it is used in the Austrian Ozone Monitoring Network to maintain and update the network as well as to distribute and transform data. For example you would like to get information about the actual measured ozone concentration of a special monitoring area: So you use your local conventions for your query as you always did, without worrying about how your neighbor measured the values belonging to the requested monitoring area. The values are presented to you as you have always been familiar with. The system itself performs for you the hard work of transforming methods and data.

The following meta information is necessary to run the Austrian Ozone Monitoring Network:

1. the operator of the local air quality measurement systems
2. the measuring stations and
3. measured components (primary and secondary pollutant, meteorological parameters, etc..)
4. formulas, averages, units and thresholds

These main categories of meta information in the Austrian Ozone Monitoring System everyone can easily retrieve by using different tools (see below).

1.2 Measured values

The measured values of one component are aggregated to and stored as half hour mean values in an environmental database. Each of these values is described by additional attributes (flags) providing information about validity and check state. Higher aggregated averages can be calculated by doing some simple definitions in the meta database.

Access to values as well as to meta information is organized by using remote procedure calls. Different implemented retrieval mechanism realize the transformation from local or remote access methods and spelling conventions into a global one. Such mechanisms exist e.g. for time series retrieval (implemented within the formula server), one for meta data updates and distribution (implemented within the meta data server), one for communication (implemented within the ozone network server), etc...

The Austrian Ozone Monitoring Network Application consists of a set of different tools for the daily work at the local authorities. These tools (discussed in (Schimak 1994c)) allow to generate daily ozone monitoring reports, to analyze time series and to visualize the current ozone situation as well as to support the handling of the austrian ozone alarming scheme.

2 THE WWW (WORLD WIDE WEB) GATEWAY

As the public interest was rapidly growing, knowing to have now valid information about the current ozone situation in Austria accessible, the Austrian Research Centre Seibersdorf in cooperation with the FEAA forced the intention to develop a WWW prototype to demonstrate how environmental information can be electronically presented to the public.

In fall 1994, after the end of the first ozone period, as it was stated that the software system is capable to work under pressure, a first release of a WWW prototype was installed at the Austrian Research Centre.

2.1 What is WWW

WWW [1, 2] (Weichselgartner, 1993 and Klute, 1994) is a new method to harmonize the different Internet services. Based on hypermedia documents and gateways for News, Gopher, Archie and FTP it was tried to simplify data retrieval in distributed systems. WWW consists of a large set of worldwide distributed hypertext documents offered by many WWW server. Each of these documents may have links to further documents or gateways for further Internet services and so on.

Access to WWW is usually performed by WWW clients. Under X Windows Mosaic is the most common user interface for WWW interactions, perhaps there are also Clients for Windows, Mac or even ASCII Terminals available. Clients providing a user interface have the great advantage that the user may not care about links. He only click on an underlined item in the document and gets data (spread all over the world) presented as he would have them locally. In terms of the Austrian Ozone Monitoring Network it means that he gets information from several air quality measurement networks.

2.2 Accessible information

In general public access (all around the world) is provided via the Seibersdorf -WWW- gateway. The server prototype itself distinguishes where the requests come from (authorized users, others) and filters the requested data depending on a special identification mechanism. The information you can retrieve is mostly the same as stored within the Austrian Ozone Monitoring Network and discussed above.

Therefore you are able to retrieve all meta data about measuring stations (list of stations) inclusive the description of each measuring station, coordinates, list of measured components as well as measured values (only checked values)

The provided information is in no way restricted to the user (accept the above mentioned privileges) thus in case of special treated data (e.g. calibration occurred during the measured period) confusion can be caused by the observer not familiar with air quality data.

Monitoring information is presented in an interactive way giving the user the ability to decide how data will be visualized. In all this presentations (Figure 1 - Figure 4) only checked and validated data are used to display ozone or meteorological data.

One of the great advantage of WWW compared with file access (FTP) is that it is possible to generate the information presentation on demand as there exists to many possibilities to offer all presentations pre-formatted. The disadvantage is, that the fetching and preparing of the requested presentation may consume time. So it gets impossible to generate for example animations (mpeg videos) on demand. There you will currently reach the upper limit of the involved computer technology. But it is possible to generate some videos in advance and give access by the WWW.

2.3 The WWW user interface

The user interface consists of some WWW pages. These pages are generated when the Seibersdorf WWW server gets a request via the gateway. The start page of the gateway to the Austria Ozone Monitoring Network is driven by the following URL (Berners-Lee, 1993)

`http://zditr1.arcs.ac.at/odv-bin/data/ui.pl`

The server is capable to decide where the request comes from, and presents the start page in german or english.

Presentation of ozone monitoring information

After having read the general information about the project you can select the kind of data presentation you are familiar with:

- two dimensional,
- three dimensional,
- colored two dimensional,
- beams about map,
- colors about map,
- mpeg-files (animations, videos)

The links *two dimensional*, *three dimensional* and *with colors* points to the map of Austria (figure 1) where you can interactively select a component from a list of pollutants and meteorological parameters as well as a time interval (for example one day / one week / one month) and an ozone monitoring station. When a station is selected by simply clicking with the mouse onto a square (representing the ozone monitoring station) within the map the values are automatically retrieved from the Austrian Ozone Monitoring Network and transformed to the requested presentation. When no station is armed with the mouse pointer (clicking between the stations) a list of stations sorted by distance from the mouse pointer is displayed to the user, where he may select stations by name.

The link *Beams (meters) about map* refers to a selection of one or more components. Selecting this presentation type the actually needed values are fetched simultaneously from

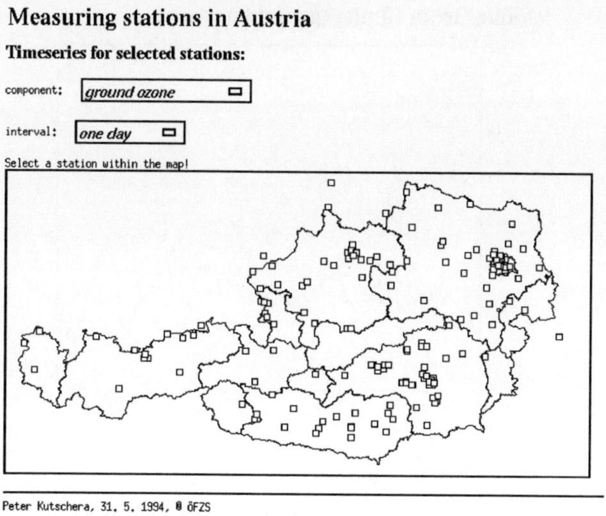

Figure 1 Selection of a monitoring station

distinct servers across the network. As there are all (10 at the moment) ozone monitoring network servers involved and information from every measuring station within the network is fetched this may cause some delay, depending on network overload.

The last link on the start page refers to a directory containing pre-calculated animations (mpeg videos).

Two dimensional data presentation: In figure 2 the two dimensional presentation of time series of one component and one station is displayed. This is the only data presentation where you are allowed to interpret single values. All the other presentations show only an overview to the trends.

There is a blue line (threshold) in the graph, which can't be seen in the hardcopy presenting the value 200*ppb*. Above this value some actions (alarm scheme) will be carried out by the Austrian government. At the end of the month (27*th*, 28*th*) you can see the typical up and down of ground ozone values within on day.

If slight changes occur from one day to the other this can't be analyzed with this kind of data presentation. Therefore use the data presentations described in the next paragraphs.

Three dimensional data presentation: In figure 3 the three dimensional presentation of a time series of one component and one station is shown. The hours within one day are drawn from left to right to show the changes of the values within one day. On the left side the days are drawn to show the changes from one day to the next.

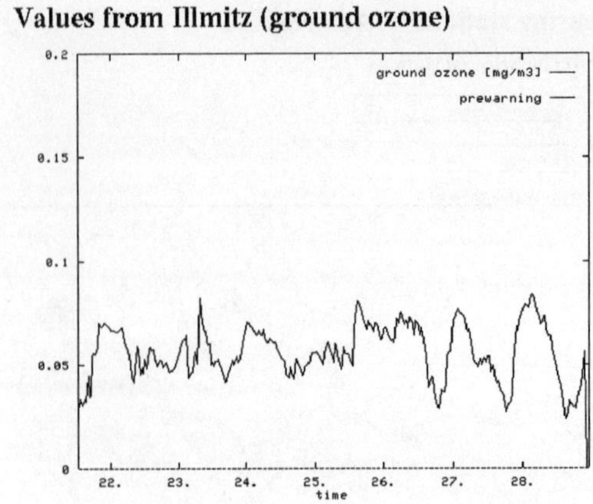

Figure 2 two dimensional presentation of time series

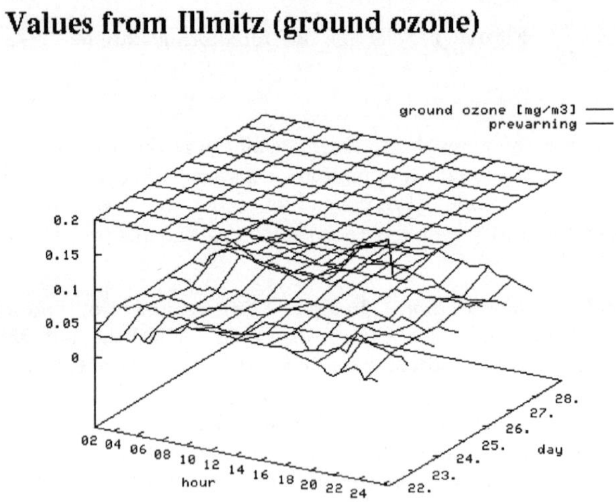

Figure 3 three dimensional presentation of time series

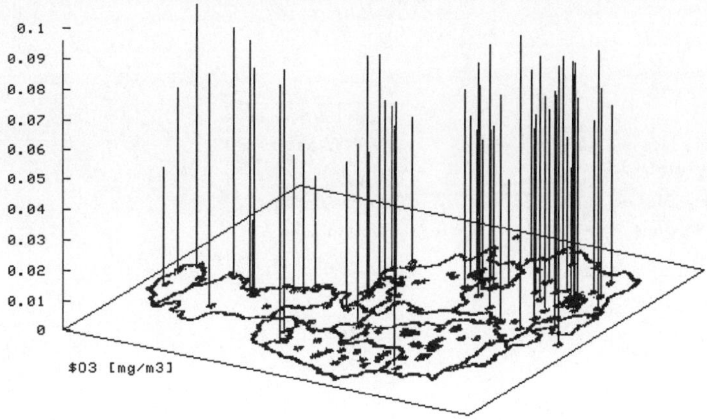

Figure 4 Beams about map

Colored data presentation: This presentation is more or less equal to the three dimensional presentation. Hours within one day are displayed from left to right to show the changes of the values within one day. Days are scaled from top to bottom.

The values are color-indicated. Its a kind of data presentation that lets you really well recognize daily (like rush hours in the morning, etc...) or long term trends (distribution of a pollutant concentration during an ozone period).

Beams about map: After selecting one or more components the figure 4 is shown. Here the values for one component (more would be possible) of all currently reachable stations are shown, but only for one time stamp. The values are shown as beams. It's not possible to get the exact values from this picture but it is useful to get a quick overview about the current ozone situation.

Animations (Mpeg videos): At the moment some pre-calculated mpeg videos can be accessed, but it is really difficult to make a hardcopy. So please have a look to our WWW pages.

2.4 Brief technical details

The WWW gateway was implemented using perl (Wall 1991 and Wall 1993) scripts. Perl is a special kind of script language, which is nearly available on each operating system. Graphical output was generated using "gnuplot" and "ppm+".

The Austrian Ozone Network Server is currently installed on a DECStation 5000/125. The operating system is ULTRIX. Remote as well as local data access is done by RPCs (remote procedure calls).

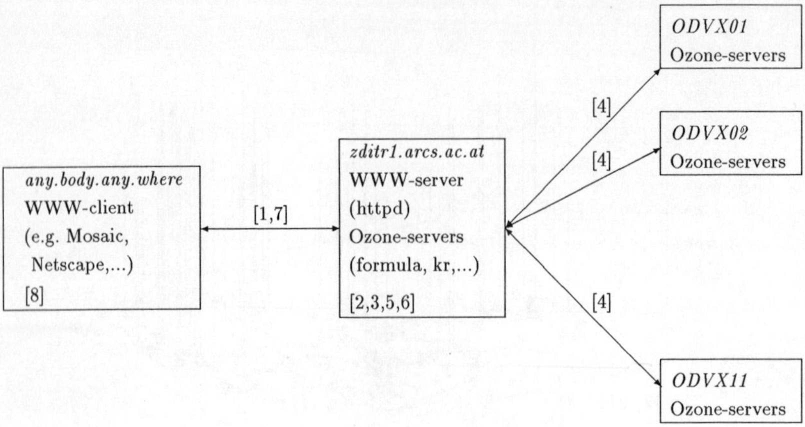

Figure 5 Server and clients

Figure 5 demonstrates the involved client and server applications used to retrieve actual ozone information. The numbers within figure 5 indicates the following actions:

1. WWW client requests URL from WWW server
2. WWW server (httpd) calls script (perl)
3. script queries information from local ozone server
4. local ozone server queries measured data from the other ozone servers in the Austrian Ozone Measurement Network
5. script generates picture (gif)
6. script generates result page (html)
7. WWW server sends result page
8. WWW client displays result

The servers in the ozone network are described in (Schimak 1994a).

The generation of the pictures, especially if more than one ozone server is involved, needs some minutes since most of the servers can only be reached which 2400*bps*. If other communication takes place it might take a little bit longer.

3 FUTURE ASPECTS

Beside the yet existing data presentation possibilities shown in figure 1 up to figure 4, for the future the following domains should be graphically displayed (Schimak 1994c):

1. ozone alarm

2. different graphical data presentation of time series for a single component and measuring station
3. combination of several components.
4. different presentations (numbers, colors, beams) of values from all stations in a map (no time series).
5. interpolated ozone values about the austrian area.
6. the information from 4 and 5 as animation

Currently a simulation and interpolation model is integrated into the Austrian Ozone Monitoring System. A future aspect may be to link the output of this model to the WWW application to present a high sophisticated model to the public.

4 CONCLUSION

The gateway to the WWW can be used to give many people access to the environmental information. We could demonstrate that the realization of a usable prototype is easy and cheap and that the WWW is a practical tool to provide environmental information as well as to develop simple front-end visualization applications.

5 BIOGRAPHY

Information about the authors can be found on

```
http://zditr1.arcs.ac.at/~peter
http://zditr1.arcs.ac.at/~schimak
http://zditr1.arcs.ac.at/~humer
```

REFERENCES

Erich Weichselgartner (1993) Unendliche Tiefen: WWW: Hypertext-basiertes Informationswerkzeug im Internet, iX 12/93, S. 84ff.

Rainer Klute (1994) Zusammengewebt: Internet-Informationsdienste mit Mosaic, iX 2/94, S. 150ff.

Tim Berners-Lee (1993): Uniform Resource Locators; A Unifying Syntax for the Expression of Names and Addresses of Objects in the Network, Internet Draft, 14. October 1993,
ftp://info.cern.ch/pub/www/doc/url7a.ps

Schimak G.(1994a), Denzer R., Humer H.: Integration in the Austrian Ozone Network, 2nd Workshop on the Integration of Environmental Data (Integration von Umweltdaten), Dagstuhl '94, R.Guettler and W.Geiger (eds.), Metropolis Verlag, Marburg

Schimak G.(1994b), Humer H., Denzer R., Knappitsch E.: The Ozone Network for Austria: Technical Concept of a Distributed Environmental Information System, 8. Symposium Informatik fuer den Umweltschutz – Umweltinformatik 94, L. M. Hilty, A. Jaeschke, B. Page and A. Schwabl (eds.), Hamburg 94, Metropolis-Verlag, Marburg

Schimak G. (1994c), Denzer R., Humer H.: Interactive Visualization in the Austrian Ozone Network, 4. Workshop, Visualization of Environmental Data (Visualisierung von Umweltdaten), Dagstuhl '94, R. Denzer, R.Guettler and H. Deutsch (eds.), Metropolis Verlag, Marburg

Wall L. (1991) & Schwartz R. L.: Programming perl, O'Reilly & Associates, 1991

Wall L. (1993)& Schwartz R. L.: Learning perl, O'Reilly & Associates, 1993

18
Global Access to Environmental Information

Ralf Kramer, Tom Quellenberg
Forschungszentrum Informatik (FZI)
Haid–und–Neu–Str. 10–14, D–76131 Karlsruhe, Germany. Telephone: (+49) 721 9654-722. Fax: -709. email: `kramer|quelle@fzi.de`

Abstract
Environmental applications deal with huge numbers of knowledge domains. In order to cope with the increasing number of highly diverse data and information sources, explicit modelling of meta-data is mandatory. This paper focuses on how data in an existing meta-data model, originally developed for German authorities, can be accessed in wide-area networks. The current implementation combines navigational access methods via Mosaic and declarative access through SQL. The system architecture and query processing in that architecture are described. Lessons learned from the initial system concern the necessity for design recovery when using an existing meta-data model and experiences obtained from using the stateless World-Wide Web (WWW) environment in database applications. Looking ahead, we sketch the active information system that will support access to semantically heterogeneous information sources in the future.

Keywords
Database systems, environmental data catalogue, meta-data, world-wide web

1 INTRODUCTION

All over the world, there is an increasing awareness for the importance of our natural environment. Environmental protection shows up in environmental politics, which is supported by environmental information systems run by state authorities. Environmental questions touch upon a large number of fairly different knowledge domains. Hence, such systems have to deal with geographic information, measurement data from different application areas and other kinds of data. Due to the decentralized organisation of the different German states, it comes as no surprise, that the authorities are organized in different ways and have developed different terminologies. In order to cope with these diverging organizations and terminologies, an environmental data catalogue (UDK) has been developed to account for the global significance of environmental affairs. This meta information system is in the process of being standardized for all German states and in addition, Austria. Current implementations of the UDK basically deliver information on the data stored in other systems, or information on whom to call (e.g., via phone). In general, they do not yet support any direct access to the original data.

Besides this application pull, there is a technological push that will determine the potential for future applications of such meta information systems. Among the most important ones are global networks. Along with these global networks come user friendly public domain tools like Mosaic for the World-Wide Web, that provide convenient access to all kinds of information using a graphical user interface.

In this paper, we describe how these two approaches, namely the meta information system UDK and graphical user interfaces for the WWW, have been combined within the project GLOBUS/UDK* [Kramer 1994]. The remainder of this paper is organized as follows: In the next section we describe these two cornerstones in more detail. In Section 3, the specific requirements for GLOBUS/UDK are given, the system's current architecture is presented in some more detail, and several examples illustrate the querying of meta-data. Section 4 summarizes the lessons learned so far from combining an already existing meta information model and system with a new interface that supports access from WWW. In Section 5, we compare our approach with related work. Section 6 presents the final conclusions and an outlook on future work.

2 CORNERSTONES

2.1 Environmental Data Catalogue (UDK)

The information required to answer many environmental questions (e.g., environmental protection) is often found in different knowledge domains located on different systems. Thus, in order to find the data and the system to answer a certain question, we first have to find out *which* information is accessible and *where* and *how* to find it. That is, we need some meta-data. Hence, a meta information system is required that consists of two main parts, namely the meta-data itself and an application program that accesses the meta-data.

The environmental data catalogue UDK and the corresponding UDK application program is a meta information system that serves three main purposes [Schütz 1994]: It should include information on all data that are relevant for environmental systems in order to be able to deliver complete answers; it should describe the quality of the original data and it should support the access to the underlying original data.

The UDK application program currently used (e.g., at the Landesanstalt für Umweltschutz (LfU), Karlsruhe, and other governmental agencies) stores its data in a relational database system. At the user front-end, the user first chooses either a catalogue that reflects the organization of the authorities in his/her specific state or country, or a catalogue that reflects certain subject-specific organizations, e.g., all data concerning air, soil or water (we will come back to these access paths in Section 3.3). Then, (s)he can reach into the meta-data directly or via additional search terms that are attached to the meta-data via something like a thesaurus. As a result of such queries, the meta-data found are presented. Since, very little of the underlying original data is accessible online today, addresses and telephone numbers are the prime means to obtain the desired information 'offline'.

*GLOBUS/UDK was funded by Umweltministerium Baden-Württemberg.

2.2 World-Wide Web and its Tools

The World-Wide Web (WWW) [Berners-Lee 1993b] is a hypertext- and multimedia-system for documents including text, pictures, movies or audioclips. Information resides on servers that are accessible, e.g., via the Internet using suitable viewers (e.g., Mosaic [NCSA 1994], NetScape). The only structuring mechanism for sets of documents is the facility to place links (Universal Resource Locators or URL) in documents leading to other documents. They consist of a protocol specification, the server address and the location of the document at the server. By following such a link, the information referenced is transfered from the server to the client and interpreted by the viewer.

Most documents on the WWW are written as HTML documents using the hypertext markup language (HTML) [Berners-Lee 1993a]. However, many client programs are able to understand further formats (e.g., plain text, postscript). The HTML format allows a structural description of text elements (headings, sections), text styles (bold, italic), and the inclusion of pictures within the text. Links to other documents can connect to text or pictures. An extension of this format (HTML+) allows the description of forms, with fields for editing text and buttons. The WWW client displays this form, the user fills it with information and then this information is sent back to the server. At the server, a specified program for this form is called via the Common Gateway Interface (CGI) [McCool 1994], which interprets the information. The output of this program is made available by the server to clients as a new HTML document. Thereby, hypertext pages can be generated dynamically, which allows a very flexible generation and presentation of information. The programs started via CGI can be written in any language available at the server (e.g., shell-scripts, perl, C, C++, ...).

With the great success and the huge number of WWW servers all around the world, some limitations and problems of the current system become apparent. One problem is the lack of access control to information available on a server: Information offered via a WWW server can be read by everyone on the net. It is not yet possible to limit page access to a specified group of people. Another problem is the lack of any structuring elements for the hypertext net; there is only a flat net of pages and links. Furthermore, these links are unidirectional only. A link cannot be updated if the page it refers to is changed or moved to another location.

3 USING WWW TO ACCESS ENVIRONMENTAL META-DATA

3.1 Requirements

The requirements for GLOBUS/UDK can be summarized as follows:

1. The modelling of UDK meta-data remains unchanged.
2. We assume that UDK meta-data is centralized both logically and physically.
3. Access across the wide-area network is read-only.
4. Adding new UDK meta-data and updating existing one is left to the UDK application program that runs on a PC. This program can be used for local read-accesses as well.
5. The functionality across the net is similar to the one of the existing UDK program.
6. The system architecture has to be based on WWW communication and tools.

3.2 System Architecture

Given the cornerstones and requirements as presented in Sections 2 and 3.1, an initial system had to be designed and implemented that allows access to UDK data stored in a relational database system across a wide-area network. This data should be accessible using Mosaic clients and the other WWW techniques described in Section 2.2.

For storing meta-data that is accessible via Mosaic, there are two alternatives. Either, data can be extracted from the database system periodically to build static HTML pages, or the database can be queried on a user's request and subsequently, the HTML pages are built dynamically.

The main problem with the first alternative is that Mosaic only supports navigational access and a full text search on the current page, which obviously is not sufficient. From the user's perspective, the second alternative offers several advantages such as declarative querying and access to the most up-to-date data. Thus, the second alternative is chosen.

Figure 1 gives a more detailed outline of the current system. The WWW client (i.e., Mosaic) accesses the WWW server across the network, which, unfortunately is *not* yet the Internet; hence, we cannot provide the URL here. The WWW server provides a single static HTML page named *start_udk* as the entry point. For all subsequent UDK accesses, the WWW server uses the CGI as described in Section 2.2 and a database precompiler program written in C. Approaches like GSQL [Ng 1993] do not yet provide the means to implement state-of-the-art user interfaces (see section 5.2). Thus, we rely on the precompiler approach, although such a solution is difficult to debug and to maintain. This program named *udk_access* retrieves the relevant data from the relational database system (Oracle Version 7) and builds the resulting HTML pages dynamically. These pages are presented to the user via Mosaic.

3.3 Querying Meta-data

In this section, we present examples for simple and complex queries that access the meta-data. To facilitate readers that are unfamiliar with German, the description of the search forms are restricted to those found in figure 2, which are easier to understand. First, let's have a look at how the meta-data is organized in the UDK.

All UDK meta-data is organized hierachically. On the first level, different countries like Germany and Austria are distinguished, on the second one, the user chooses a state within a country, e.g., Baden-Württemberg. Finally, for each state, there are two catalogues to choose from: the first one reflects each state's organization of authorities, the second one organizes the meta-data according to their content.

In the middle part of the left screendump shown in figure 2, we see the user's initial choice among two catalogues named *UDK* and *GDK* for the states Baden-Württemberg (*BW*) and Niedersachsen (*NI*) located in Germany (*GER*). Having made the initial choice among the catalogues available, the user may search for addresses or so-called *UDK-objects*. Simple searches can be used to locate addresses or *UDK-objects*, while more complex searches can only be used with *UDK-objects*. An example of a complex search is found in figure 2.

For simple, straight-forward queries, the search parameters, which may include the usual SQL wildcards, are mapped to several attributes of different relations within the precompiler program *udk_access*. This hides the complexity of the underlying data model from

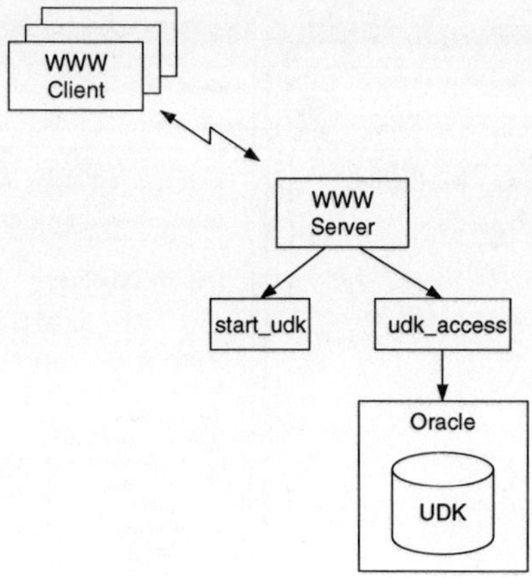

Figure 1 A detailed Look of the current System

the user. The query result is presented in a different form that is generated dynamically depending on the result tuples.

The right part of figure 2 depicts an example of a more complex search. Basically, each of the 7 search attributes (i.e., *Objektname, Adresskennzeichen, ...*) target several attributes of a single relation. In combination with the search parameters, the user can choose among a set of comparison operators (e.g., $=, \leq, <, \ldots$). The triples *search attribute*, *comparison operator*, *search parameter* are combined using intersection and union operations (i.e., *AND* and *OR*). Obviously, this flexible and powerful mechanism requires dynamic SQL within the program *udk_access* and hence, a more elaborate implementation than the simple searches sketched above. However, the underlying complexity remains completely hidden from the user.

As a result of both simple and complex queries, the user first receives a list that includes important information relating to addresses and UDK-objects. In these lists, the user can choose those addresses and objects from which (s)he wants to obtain further details. Then, as the second result, all relevant attributes from the underlying relations are presented in another form that has been generated dynamically.

4 LESSONS LEARNED

In this section, we discuss the most important lessons obtained from the initial system described in Section 3. The lessons include the impact of poor database design on the implementation effort, deficiencies of the current WWW tools for implementing state-of-the-art database user interfaces, and the impact of using a stateless communication paradigm.

In contrast to many other environmental (meta-) information systems we started with an already existing meta-data model. The starting point of our project were some 15 base relations that comprise the UDK's most important data. Unfortunately, only a non-

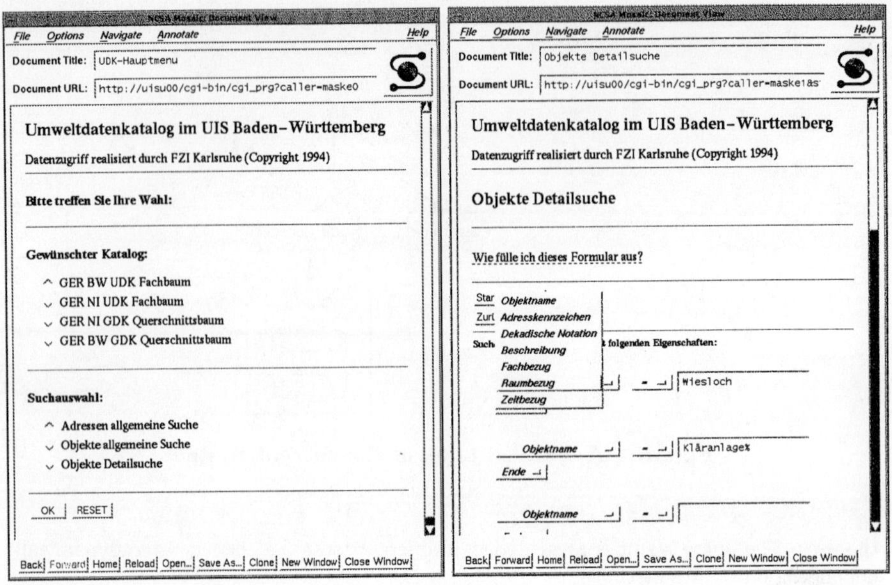

Figure 2 Choosing among different Data Catalogues and Complex UDK-Object Search

standard graphical diagram and the tables (for the relational database system Oracle, Version 7) were available. Some pecularities of the database design (e.g., lack of data independency) were hard to understand by simply examining the schema, i.e., without looking at the data. Thus, we used the UDK's PC implementation to run against the same meta-data as in our implementation in order to find the "correct" SQL queries. That is, those queries that yield the same results in our application as in the PC implementation. In our case, the existing meta-data model required a design recovery [Chikofski 1990] (usually refered to as the first step of reverse engineering process) in order to implement the new WWW/Mosaic application.

As already pointed out at the end of Section 3.3, current WWW tools require a pre-compiler to implement state-of-the-art user interfaces. That is, interfaces in which the format of query results do *not* match the search format. With the precompiler approach, even minor changes in the database design or in the generated HTML pages require a substantial knowledge of the implementation. Thus, a higher abstraction level and some sort of interpretation or precompilation to the lower level are mandatory to achieve a better maintainability.

Different from traditional database applications (stand-alone and client-server applications), communication in WWW/Mosaic environments is stateless. Thus, there is no notion of a session or (ACID-) transaction across the net, i.e., when a WWW client (like Mosaic) demands a HTML document from a WWW server. Classical database transactions only take place locally at the server's site within the precompiler application (see figure 1). We agree with the conclusions of [Eichmann 1994] that – as a side-effect – user interfaces are likely to become more simplistic. For example, the move to a stateless environment limits the number of actions a window or page is able to process. Approaches like the one taken in Zelig [Varela 1994] to generate database applications dynamically are another promising approach to allow for adaptable user interfaces across the net.

5 RELATED WORK

In this section, we place our work in the context of other environmental information systems (Section 5.1) and – independent from a specific application area – in the context of database access in wide-area networks (Section 5.2). The list of approaches, concepts, and systems discussed does not claim to encompass all relevant work in the area. We restrict the discussion to those that we think to bear the most direct relationships to our problem and solution.

5.1 Environmental Information Systems

In this section, we discuss those environmental information systems that support – or, to be more precise, at least intend to support – the access to environmental data in wide-area networks. In contrast to most other approaches, we employ an already existing and fairly elaborate meta-data model.

Up to now, only one approach to make the UDK available across the net has been published [Lessing 1994]. [Lessing 1994] mainly discusses issues that are complimentary to our focus, namely security and authorization problems when accessing the UDK and the underlying original data. In contrast, we concentrated on making the UDK's meta-data available using WWW tools. Hence, we have not yet addressed the security issues mentioned in section 2.2. We feel it worthwhile to examine more closely how to incorporate the approach [Lessing 1994] in our system.

The CERES Global Knowledge Network (CERES-GKN) [CERC 1994] aims at supporting the environmental-conscious design of products [Evans 1994]. The intended user communities are – among others – companies and their engineers, and not authorities as in our implementation. While the users are different, the first implementation of CERES-GKN is likely to be based on WWW tools as well. So, although the intended audiences are different, the techniques used or intended to be used, are similar with ours.

The Consortium for International Earth Science Information Network (CIESIN) aims easing the access to information and data, used in facilitating the understanding of human interactions with the environment [Zimmermann 1994]. The intended audience for this application are mainly scientists and the public. As mentioned previously, CERES-GKN focuses on decision making in business and industry and GLOBUS/UDK – at least in the short run – on providing information for authorities. CIESIN offers both access to data and meta-data.

According to [Crossley 1994], the Australian Environmental Resources Information Network ERIN [ERIN 1994] uses Wide Area Informations Server WAIS to access environmental data. WAIS server use indexes for each word in every document; the index is used for query processing. In [Crossley 1994], the importance of meta-data and advantages of using database systems to manage the meta-data are discussed. With ERIN's focus on spatial queries, a relational database system is used to store both data and meta-data

[Boston 1994]. The system architecture is very similar to the one we use (see figure 1). Just like in our approach (see section 3.3), ERIN queries first deliver a summary report before the user can select further information on individual records [Boston 1994, Section 2].

5.2 Database Access in Wide-Area Networks

For database access via WWW-servers, systems like GSQL [Ng 1993] support the generation of forms and queries. Unfortunately, they offer only a basic functionality for the presentation of the results. In particular, it is impossible to use results of queries in detailed query forms or to combine navigational and declarative access flexibly. To summarize, although such tools provide a higher abstraction level, they do not meet the requirements of today's user-interfaces like the one sketched in section 3.3 and shown in figure 2.

In the Repository Based Software Engineering (RBSE) project, a repository based on meta-data is accessible via Mosaic and the WWW. Similar to our system, the current MORE system [Eichmann 1994] is a reimplementation of a 'conventional' database application. Thus, the lessons learned are similar to ours, with the notable exception that we did not reengineer the database design.

From a database system point of view, several architectural approaches such as, distributed databases, multi-databases, and federated databases, have been proposed to solve the problem of heterogeneity at technical, syntactical, and semantical levels. We refer to [Litwin 1990, Sheth 1990] for an overview. With its present lack of any active capabilities, our approach can be characterized as a federation, i.e., a number of loosely coupled systems that retain their autonomy while participating in the federation. Unfortunately, federated systems are unable to solve the problem of semantic heterogeneity [Papazoglou 1992]. Thus, a certain degree of activity will be required to achieve the desired interoperability.

6 CONCLUSIONS AND FUTURE WORK

In this paper, we have presented the approach taken in GLOBUS/UDK to access environmental meta-data in wide area networks. Demonstrations and first user tests of our system showed that we have implemented a viable approach combining navigational and declarative accesses using technology that is currently available. The meta-data is stored in a relational database system. The application itself, allows for a combination of navigational and declarative access by combining the capabilities of Mosaic and SQL.

The vision or goal of the final system can be described as an active information system that supports access to semantically heterogeneous information sources in multiple user-specific terminologies. The current UDK serves as the core and starting point for a - logically - centralized general meta-data repository. Currently, so-called intelligent and cooperating information systems [Papazoglou 1992] are viewed as the most promising approach to cope with the problems of system federations. We plan to address the severe problem of semantic heterogeneity and semantic interoperability [Jamil 1994] in the future.

Acknowledgements: The students K. Ateyeh, O. Scheffczyk, and M. Walz were instrumental in implementing GLOBUS/UDK. We gratefully acknowledge their support. Special thanks are due to the other members of FZI's and the University of Karlsruhe's Information Systems Group joint working group on system federations (Dr. G. von Bültzingsloewen, A. Koschel, C. Reck, B. König-Ries, R. Schmidt, D. Theobald, and Dr. M. Wallrath) for the stimulating discussions. Their suggestions and Prof. Dr. P. C. Lockemann's comments on an earlier version of this paper significantly improved its focus.

REFERENCES

[Berners-Lee 1993a] T. Berners-Lee. Hypertext mark-up language. Technical report, CERN, http://info.cern.ch/hypertext/WWW/MarUp/MarkUp.html, 1993.

[Berners-Lee 1993b] T. Berners-Lee. The world wide web initiative. Technical report, CERN, http://info.cern.ch/hypertext/WWW/Theproject.html, 1993.

[Boston 1994] Tony Boston and David Stockwell. Interactive species distribution reporting, mapping and modelling using the World Wide Web. In *Second International World Wide Web (WWW) Conference '94: Mosaic and the Web*, Chicago, USA, October 1994. http://kaos.erin.gov.au/database/WWW-Fall94/species_paper.html.

[Chikofski 1990] E.J. Chikofski and J.H. Cross II. Reverse Engineering and Design Recovery: A Taxonomy. *IEEE Software*, 7(1):13–18, January 1990.

[CERC 1994] The Concurrent Engineering Research Center, West Virginia University, USA; The CERTIA Research Center, University of Roma at Torvergata, Italy; The Advanced Research Laboratory, Pennsylvania State University, USA. *Inaugural Workshop for the CERES Global Knowledge Network*, Berkley Springs, West Virginia, USA, April 1994. http://www.cerc.wvu.edu/ceres/inaugural_workshop_index.html.

[Crossley 1994] David Crossley. WAIS throug the Web Discovering Environmental Information. In *Second International World Wide Web (WWW) Conference '94: Mosaic and the Web*, Chicago, USA, October 1994. http://www.ncsa.uiuc.edu/SDG/IT94/-Proceedings/Searching/crossley/paper.html.

[Eichmann 1994] D. Eichmann, T. McGregor, and D. Danley. Integrating Structured Databases Into the Web: The MORE System. In *First International Conference on the World-Wide Web*, Geneva, Switzerland, May 1994. CERN. Preliminary Proceedings, http://www1.cern.ch/WWW94/PrelimProcs.html.

[Elmagarmid 1990] Ahmed K. Elmagarmid and Calton Pu. Special issue on heterogeneous databases. *ACM Computing Surveys*, 22(3), September 1990.

[Evans 1994] S. Evans and E.P. Rose. Environmentally Conscious Design Using a Global Knowledge Network. Technical report, Cranfield University, CIM Institute, UK, 1994.

[ERIN 1994] ERIN General Information. http://kaos.erin.gov.au/general/general.html.

[Jamil 1994] Hasan Jamil and Paul Johannesson. Sematic interoperability - context, issues and research directions. In Michael Brodie, Mathias Jarke, and Michael Papazoglou, editors, *Proc. of the 2nd Intl. Conference on Cooperative Information Systems*, pages 180–191, Toronto, Canada, May 1994.

[Kramer 1994] R. Kramer and T. Quellenberg. Global Access to Environmental Information. FZI Report 9/94, Forschungszentrum Informatik (FZI), Karlsruhe, Germany, December 1994.

[Litwin 1990] W. Litwin, L. Mark, and N. Roussopoulos. Interoperability of multiple autonomous databases. In *ACM Computing Surveys* [Elmagarmid 1990], pages 267–293.

[Lessing 1994] H. Lessing and T. Schütz. Der Umwelt-Datenkatalog als Instrument zur Steuerung von Informationsflüssen. In L.M. Hilty, A. Jaeschke, B. Page, and A. Schwabl, editors, *Informatik für den Umweltschutz; 8. Symposium, Hamburg 1994; Band I*, Umwelt-Informatik Aktuell, pages 159–167, Marburg, 1994. Metropolis.

[McCool 1994] Rob McCool. Common gateway interface overview. Technical report, National Center for Supercomputiong Applications (NCSA), http://-hoohoo.ncsa.uiuc.edu/cgi/overview.html, 1994.

[NCSA 1994] NCSA. Introduction to NCSA Mosaic for X. Technical report, National Center for Supercomputing Applications (NCSA), http://www.ncsa.uiuc.edu/SDG/-Software/Mosaic/Docs/d2-intro.html, 1994.

[Ng 1993] Jason Ng. GSQL: A Mosaic-SQL gateway. Technical report, National Center for Supercomputing Applications (NCSA), http://www.ncsa.uiuc.edu:80/SDG/-People/jason/pub/gsql/, 1993.

[Papazoglou 1992] Mike P. Papazoglou, Steven C. Laufmann, and Timos K. Sellis. An organizational framework for cooperating intelligent information systems. *Intl. Journal on Intelligent and Cooperative Information Systems*, 1(1):169–202, March 1992.

[Schütz 1994] T. Schütz and R. Böhm. Die Datenstrukturierung des Metainformationssystems Umwelt-Datenkatalog. In H. Kremers, editor, *Umweltdatenbanken*, number 5 in Praxis der Umweltinformatik, pages 245–258, Marburg, 1994. Metropolis.

[Sheth 1990] Amit P. Sheth and James A. Larson. Federated database systems for managing distributed, heterogenous, and autonomous databases. In *ACM Computing Surveys* [Elmagarmid 1990], pages 183–236.

[Varela 1994] C.A Varela and C.C. Hayes. Zelig: Schema-Based Generation of Soft WWW Database Applications. In *First International Conference on the World-Wide Web*, Geneva, Switzerland, May 1994. CERN. Preliminary Proceedings, http://-www1.cern.ch/WWW94/PrelimProcs.html.

[Zimmermann 1994] F. Zimmermann and P. Young. Information for a Changing World: CIESIN's Use of WWW to Facilitate Access to Data and Information Relevant to Human Interactions in The Environment. In *Second International WWW Conference '94*, Chicago, USA, October 1994. http://www.ncsa.uiuc.edu/SDG/IT94/Proceedings/-EarSci/zimmerman/% -zimmerman.html.

BIOGRAPHY

Ralf Kramer is project manager and Executive Vice-Director of the database research group at FZI, Karlsruhe. He holds a Diploma degree (Dipl.-Wi.-Ing.) and a Ph.D. (Dr. rer. nat.) from the University of Karlsruhe. His main research interests are databases in distributed, heterogeneous information systems, both from a technical and from an application's point of view.

Tom Quellenberg works as a research assistant in the database research group at FZI, Karlsruhe. He holds a Diploma degree (Dipl. Inform.) from the University of Bremen and is mainly interested in workflow management and object oriented database systems.

SIRIUS - Saarbrücken Information Retrieval and Interchange Utility Set

Reiner GÜTTLER and Ralf DENZER
Hochschule für Technik und Wirtschaft des Saarlandes
Goebenstr. 40, D-66117 Saarbrücken, Germany, tel. 49-681-5867-238 or 426, fax 49-681-5867-122, email guettler@htw.uni-sb.de or denzer@htw.uni-sb.de

Abstract

The SIRIUS-project deals with the development of a prototype of a tool set for the integration of heterogenous subsystems fournishing environmental information. We first discuss the major integration problems which delay the development of Environmental Information Systems (EIS). The objectives of the SIRIUS-project and its integration concept are presented. Finally we describe the implementation of the prototype and its functionalities.

Keywords

Environmental Information Systems (EIS), integration, open distributed systems, meta-information

1 PROBLEMS IN THE FIELD OF EIS

Since the midth of the 80th we observe an increasing use of computer based methods in dealing with environmental problems. There are two main reasons :
an increasing pressure on environmental authorities by the public ending in additional laws and an increasing state of knowledge In the field. Both reasons lead to an enormously increasing amount and complexity of work for those authorities which they can no longer overcome without computer based methods.
Typical aspects of environmental computer applications are often the interdisciplinarity and the need for data and methods coming from different sources. This leads to the main problem of integration of environmental data and methods into an Environmental Information System (EIS).
One approach could be a centralized solution with a top-down designed overall system to avoid heterogenous subsystems. Many disadvantages of such an approach have soon become obvious, among others :

- specific objectives of different subsystems need different hard-/software-solutions
- already existing subsystems cannot be integrated

We state that it is impossible for overall systems of such a size

- to find suitable concepts
- to organize and coordinate such systems
- to maintain such systems

under the condition that every user has a solution which fits his specific needs.

This leads to the alternative idea of a completely decentralized open distributed EIS, in which we „can live with heterogenous subsystems". Subsystems can be heterogenous in different aspects - syntax, semantics and structure - which we will explain shortly.

Heterogenous **syntax** means that the subsystems can differ in hardware, operating system, data base, local network, communication and user-interface software. Since the typical users (e.g. in authorities) in general will not have a high qualification in dealing with computers, heterogenous syntax of subsystems must be hidden for those users. With the increasing efficiency of modern software-tools this kind of heterogeneity should no longer constitute an obstacle too difficult to overcome.

Heterogenous **semantics** seems much more difficult in our opinion. We understand by heterogenous semantics different comprehension of the meaning of an information. This starts by differing meanings of a single information (e.g. different names for the same object, the same name for different objects) up to different association of additional information to a specific object. There are many aspects of heterogenous semantics (see Denzer,1994a for a detailed explanation).

Heterogenous **structure** means the use of different data structures for the same objects in different subsystems. Since already the raw data can be heterogenous in syntax and semantics, the structures will be heterogenous by this fact, but a new type of heterogeneity arises by the different aggregations and components of raw data to built structures.

By the use of meta-data the problems arising from the different kinds of heterogeneity can be documented but in general **not** be removed. We can for example state by the use of meta-data that two pieces of information that we want to compare are not comparable because of different measurement methods, different aggregations and so on. There are only few cases, where computer based methods can help in this situation, for example by conversion of data following a standard framework. The reason for the problem lies much deeper. The different scientific disciplines in the field (biologie, chemistry and so on) should be aware, that computer based methods in general cannot heal a lack of strategie in the production of raw data.

The **dynamic** behaviour of the different subsystems constitutes another difficult problem for integration. The scientific methods in the field of ecology are still young. So it is inevitable that the subsystems must be able to develop independently with the increasing knowledge in the field.

Still another problem arises from the **autonomy** of the different authorities that are often responsable for the different subsystems. In many cases an institution (e.g. a ministry) that needs data from a subsystem does not have the right to give orders to the institution responsable for the subsystem (depending from another ministry).

All those problems constitute at once arguments against a centralized approach but also quite difficult obstacles for a distributed solution. They are described in more detail in GÜTTLER, 1993, DENZER, 1994a and DENZER, 1995.

2 OBJECTIVES OF THE SIRIUS-PROJECT

The main objectives of the SIRIUS-project were the conception and a prototype for an open distributed system combining different heterogenous subsystems including the necessary meta-informations.

The solution we were looking for should satisfy the following requirements :

a) The solution should allow to integrate subsystems which are heterogenous in any aspect (syntax, semantics, structure). This means that the system can work without any structure of meta-data or raw-data imposed by any organisation. Obviously it is possible for the system to work with an imposed structure by a ministry for example. But it is important that the system is not the restricting factor. So the system can adapt to the problem, e.g. organisation needs, and not vice-versa.

b) Any client should be able to investigate any subsystem for meta-data and the raw-data behind with one single tool. It should also be possible to transport the needed data through the network.

c) The solution should allow any subsystem to modify dynamically. This means that neither any client neither the investigation tool need to be modified, if a subsystem changes Its structure. Next time a client accesses the modified subsystem he will notice the modification.

3 THE INTEGRATION CONCEPT

The requirements lead to some concepts based on ideas of several earlier projects (DENZER, 1993, SCHIMAK, 1994). From DENZER, 1993 comes a two-step model for an open EIS satisfying point a) and b) of the requirements. The model tries to seperate the components of the local systems from the network by a kind of abstraction realized by a layer between local systems and network, the components of this layer beeing some public services. The local systems treat their data using local conventions and following local needs. They are designed for local purposes and very often exist already when the problem arises to let them communicate through a network. This means that the design of the local systems very often did not take into account the fact to open the system for external access. The local systems have an interface with the components of the intermediate layer (S1), while these components communicate via a standard protocol (S0) through the network.

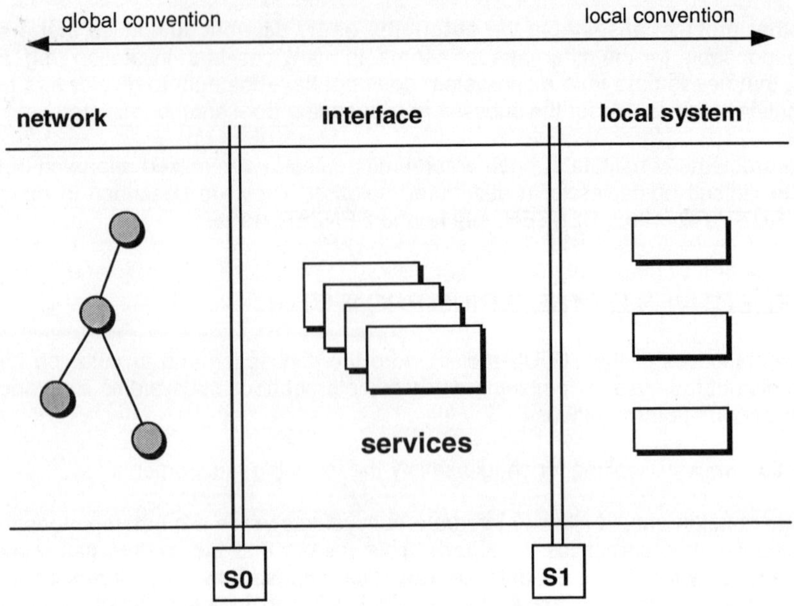

A local object interface (S1) is used for the local communication of the subsystems with the services available on every node of the network . A general object interface (S0) is used for the communication of the different servers. This general object interface must be publicly agreed upon.

A successfull implementation of the basic idea of this concept is described in SCHIMAK, 1994. But this implementation does not yet satisfy the requirement c). A client must know the semantics and structure of an object fournished by any server.

But based on experiences in the above cited implementation an additional idea (see also DENZER, 1994b) was integrated in the SIRIUS-concept : the abstract object interface. It means that any subsystem should be able to describe its content over an interface which is uniform in the whole system. The interface must be able to communicate any information types with their relations. Since every information communicates in addition its structure the clients are not sensible for modifications and thus the requirement c) is satisfied.

Example : it is possible to implement an Eenvironment Data Catalogue (EDC) - visualization-tool in such a way, that a client must not be modified even if

- the structure of the EDC is modified dynamically
- object classes in the EDC are modified dynamically
- at different nodes are different EDC's

4 IMPLEMENTATION : THE SIRIUS-TOOLKIT

We give first a list of the SIRIUS components and describe below how these components can work together to match our concepts.

The main SIRIUS-components are:

- interface descriptions of S0 and S1 in forms of abstract data types
- an EDC Integration Tool
- an EDC Cat Server : a network-wide service offering a local meta-information sytem for remote access (S0-interface)
- a EDC visualization tool : network-wide service for the visualization of local or remote EDC-informations

Besides these main components there are some C-programmed interfaces helping to use local tools to access the communication interfaces or to use SIRIUS user-interface-objects for local applications.

Following the SIRIUS-concepts the implementation allows the different nodes to access the SIRIUS-system as an information server at different levels.

SIRIUS offers the subsystems an EDC Integration Tool, a network service to combine distributed EDC's. This tool lets two options for a subsystem to be integrated as an information server into SIRIUS :

- Either the subsystem disposes already of a meta-data-system or it decides to built such a meta-data-system independently within the system or in a specific meta-information-system. In this case the subsystem needs only an additional EDC Local Server in front of his subsystem, which offers the meta-data using the S1-interface to our EDC Integration Tool. In general no modifications will be necessary for the rest of the system.

- If there is no meta-information-system in the local subsystem, the EDC Integration Tool can be used as an interactive comfortable tool to built a local meta-data-system. It is important to state that there are only a small set of functions to be implemented and these functions are rather easy tasks. There are several advantages choosing this alternative: 1) integration is possible very quickly and at low cost; 2) the catalogue and class structure presented to the network may be different from the structures used in the local system; 3) the local system remains completely untouched.

A SIRIUS-user at one subsystem can use the EDC Visualization Tool (browser) or any self-made tool which is able to access the information using the SIRIUS-interfaces to visualize any local or remote EDC-information. He accesses the network-wide EDC Cat Server which communicates with the Cat Server implementations at other network nodes via S0 to communicate the desired informations to the user. Every Cat Server implementation accesses local Meta-informations via S1. The Cat Server knows if there is a EDC Local Server in the subsystem or if the meta-information is found in a EDC Local Server in the inetermediate layer (a meta-data-system built-up using the EDC Integration Tool).

The services offered by a local system can be divided into two categories :

- node independent services : these are services which have to be implemented at every node. These services can be used without any knowledge of the structures and semantics of the information in the subsystem (e.g. „show me all classes of objects in the subsystem", „show me all objects of a specific class", „give me all objects with a specific characteristic", and so on).
- node specific services (additional services) : they use the structure and the semantics of the information and thus cannot be used without any additional knowledge. Furthermore they will need in general specific implementations and thus will provoke modifications in the tools usinf these services.

The functionality offered to the user at any local subsystem can be described as follows.
First the system presents him an overview, which subsystem actually is connected and thus can be accessed, i.e. which subsystem offers a catalogue and a local meta-server. The overview is actually only a kind of navigation help and could be extended to a real navigation layer by different means, e.g. the introduction of a distributed keyword system. In any case the main aspect must be respected that the distributed system is presented to the user logically as a single unit such that the user must not be aware of the distribution.
Using the navigation tool the user can choose a node - in the case of a real navigation layer he wood not be aware of doing so - and request information. The system offers him catalogue and class information from the local database. Internally this is done by a local EDC Server already existing or built up using the EDC Integration tool as described. The user can review catalogue and classes in order to locate information which might be interesting for him. The user then can ask for specific objects which will be sent back to him.
It is possible for the user to navigate from one node to the other. When he changes from node1 to node2 and later again to node1, the data structures of node1 may have been modified considerably resulting in a modified catalogue and classes. The system will offer him the modified information. Thus a decentralized update is possible without any affect outside the local systems.

We summarize the advantages of using SIRIUS as an integration tool :
a) There is a standard abstract object model and a standard protocol which is able to transport any kind of structures information.
b) The dynamic behaviour of the local subsystems is respected, a fact that is very important in the case of environmental information. It is not necessary to exchange clients when the servers change.
c) The local integration is always done using a small set of simple functions.
d) It is possible to start with a small implementation using the interactive tool and the standard browser client to connect to the network (it is even possible to start with no local information system at all if the local information is not yet available electronically).

The SIRIUS-software is based on standard RPC software and C as programming language and thus available on a wide range of systems from personal computers up to powerfull workstations or even minicomputers or mainframes. For a local system every part of SIRIUS might be replaced, e.g. the standard browser, the integration tool or the server stubs may be used but must not be used. SIRIUS includes libraries to program the local system itself (including user interface objects) and makes it possible to tailor the local integration scheme to local requirements. Only the standard object protocol is necessary at any node in order to make the network run.

5 REFERENCES

DENZER, (1993), Architektur offener Umweltinformationssysteme, Zwischenbericht DFG-Projekt De501/1-1

GÜTTLER, DENZER, (1993),Umweltinformation als verteiltes System, in : A. JAESCHKE et al. (eds), Informatik für den Umweltschutz, 7. Symposium, Ulm, Informatik Aktuell, Springer, pp. 404-416

DENZER, GÜTTLER, (1994a), Integration von Umweltdaten, in B. PAGE et al., Umweltinformatik, Oldenbourg,

DENZER, (1994b), Object-interfaces für Umweltinformationssysteme, in L.M. HILTY et al. (eds), Informatik für den Umweltschutz, 8. Symposium, Hamburg 1994, Metropolis, pp. 343-350

DENZER, GÜTTLER, (1995), Über die Integrationsproblematik bei übergreifenden Umweltinformationssystemen, GI-SI Fachgespräch „Länderübergreifende Umweltinformationssysteme", Springer 1995 Zürich

SCHIMAK, DENZER, HUMER, KNAPPITSCH, (1994), The Ozone Network for Austria - Technical Concept of a Distributed Environmental Information System, in L.M. HILTY et al. (eds), Informatik für den Umweltschutz, 8. Symposium, Hamburg 1994, Metropolis, pp. 89-96

PART SEVEN

Artificial Intelligence Applications

20

Biological Monitoring: a Comparison between Bayesian, Neural and Machine Learning Methods of Water Quality Classification.

W. J. Walley
School of Computing, Staffordshire University
Beaconside, Stafford ST18 0DG, UK.
Tel: +1785 353510, Fax: +1785 353497
E-mail: W.J.Walley@soc.staffs.ac.uk

S. Džeroski
Institut Jožef Stefan
Jamova 39, 61111 Ljubljana, Slovenia.
Tel: +386 61177 3217, Fax: +386 61125 1038
E-mail: saso.dzeroski@ijs.si

Abstract

Biological methods of monitoring river water quality have enormous potential but this is not presently being realised owing to inadequacies in methods of data interpretation and classification. This paper describes the development and testing of several classification models based on Bayesian, neural and machine learning techniques, and compares their performance with two traditional models. It is demonstrated, using an expertly classified test data set, that 'naive' Bayesian models and multi-layered perceptrons can significantly out-perform the traditional methods. It is concluded that these two techniques presently provide the most promising means of realising the full potential of bio-monitoring, either acting separately or jointly as complementary 'experts'.

Keywords

Bio-monitoring, water quality, classification, Bayesian, neural networks, machine learning

1 INTRODUCTION

Biological Monitoring

Biological methods of monitoring environmental quality are of increasing importance. The basic principle of these methods is that the type and diversity of the biota at a given site reflect its environmental quality. The biota act as resident observers of prevailing conditions and provide evidence which can be interpreted into quality terms. Kolkwitz and Marsson (1902) were the first to propose the use of biota as a means of monitoring the quality of natural waters. Since then many different methods of mapping biological data to discrete quality classes or continuous scales have been developed (DePauw and Hawkes, 1993). For practical reasons, the majority of these methods are based exclusively on the composition of the macro-invertebrate community living on or in the bed of the river or lake. Unfortunately, the rationales behind the methods used to interpret the community data are typically *ad hoc* and highly subjective, and their reliabilities have been less than desired. Walley *et al.* (1992) and Ruck *et al.* (1993) first demonstrated the potential of Bayesian and neural classifiers to provide more reliable interpretations, and this present study extends the search for better methods via an appraisal of an even wider range of AI (Artificial Intelligence) techniques.

Basis of the Study

A comparative study of a wide range of computer-based methods of classification (Michie *et al.*, 1994) found that relative performances differed significantly between domains, depending upon the nature of the data. However, none of the data sets used in the study was in any way related to biological monitoring of environmental quality, so no conclusions can be drawn from it as to the best methods to apply in this field.

The aim of this study was to develop and test several computer-based methods of classifying river water quality with a view to identifying the method having greatest potential for future use. This was achieved by comparing their performances relative to traditional methods using test data which had been classified by an expert. The computer-based methods used three different AI techniques: 'naive' Bayesian inference (a probabilistic approach); supervised and unsupervised neural networks and machine learning of regression/model trees (Quinlan, 1993). The traditional methods used were the Average Score Per Taxon (ASPT) and Trent Biotic Index (TBI) as applied in the United Kingdom.

2 THE DATA SET

The raw data set consisted of 292 biological samples taken from sites on the Upper Trent in England. The samples had been classified in water quality terms by an acknowledged expert river ecologist, H. A. Hawkes. This gave the best possible mapping from data to quality class, given the present state of the art. The ASPT's and TBI's were derived by the National Rivers Authority and supplied with the data. Since some of the taxa (i.e. different organisms) found in the samples occurred infrequently or were of little indicator value, a short-list of 36 taxa was drawn up by the expert to form the basis of the project database. The list comprised 7 species, 2 genera, 26 families and 1 order, and was designed to provide effective coverage of the whole range of water quality. Within the samples each taxon was identified as being either: absent, present (1-2 individuals), few (2-9), common (10-49), abundant (50 - 99), very abundant (100 - 1000) or greater than 1000. For the purpose of the neural and machine learning models, these six 'states of existence' were coded in the database as -0.2, 0.2, 0.4, 0.6, 0.7, 0.8 and 0.9 respectively. Thus, the absence of a taxon, being represented by a non-zero value, was able to contribute to the classification. Water quality was represented by five exhaustive and mutually exclusive classes (B1a, B1b, B2, B3 and B4) from the best to the worst quality. These were represented by Classification Indices (CI) from 1 (B4) to 5 (B1a) in the database. However, when classifying the samples the expert was able to identify good (+), normal and bad (-) samples within each class. Thus each class was effectively sub-divided into thirds, so these were represented within the database by CI's differing by 0.33 of a class interval (e.g. B2+ = 3.33). To facilitate independent testing of the models, the database was divided into two representative sub-sets: a *training set* of 195 samples and a *test set* of 97 samples.

3 THE MODELS TESTED

Traditional Methods

The two traditional methods tested in this study, the TBI and ASPT, are typical of the many different systems which are presently in use. They are subjective *ad hoc* systems based on simple arithmetic and/or look-up tables. De Pauw and Hawkes (1993) comprehensively reviewed most of the bio-indicator systems used in Europe, including the TBI, ASPT and Saprobic systems. The TBI was first used by the Trent River Board in England in the late 1950's, and has since formed the basis of indices adapted for use in France, Belgium and Italy. The ASPT is based upon the BMWP score system which was developed in the late 1970's by the Biological Monitoring Working Party, and is unique to the UK. Unfortunately, the Saprobic system, which is used in Germany and many east European countries, could not be tested in this study because its data requirements were not compatible with the project data set.

The TBI is based on a two-way look-up table in which the rows represent different taxonomic groups arranged in decreasing order of pollution sensitivity, and the columns represent increasing levels of the total number of groups present. The user looks up an index (between 0 and 10) at the intersection of the row corresponding to the most pollution sensitive group in the sample and the column corresponding to the total number of groups in the sample. Essentially, it is a subjective pattern recognition system based upon these two features.

In the BMWP system each macro-invertebrate family has been allocated a score (between 1 and 10) indicative of its sensitivity to pollution. Thus pollution sensitive animals like stoneflies score 10, whilst pollution tolerant ones like sludge worms score 1. The BMWP score for a given site is derived by summing the scores of all the families present at the site. However, since this is sensitive to sampling effort and the seasonal behaviour of some of the families, practitioners now favour the ASPT (i.e. BMWP score divided by the number of contributing families). This, being an average, is less sensitive to sampling effort and seasonal effects and is considered the more reliable indicator.

Thus, in theory the ASPT has a continuous scale from 0 to 10, but in practice its range is from about 1.5 to 7.5, corresponding approximately to classes B4 to B1a. The equivalent range for the TBI is 1 to 10. Linear regression on the *test set* was used to map these to the 1 to 5 scale adopted in the study. The resultant mappings were: ASPT/1.28 and TBI/2.00.

Naive Bayes

When Bayesian inference is applied to classification problems involving several mutually exclusive and exhaustive classes it is normal to assume that all items of evidence are conditionally independent. Such models are commonly referred to as 'naive' Bayes, because the assumption is considered naive. However, in bio-monitoring the assumption that the occurrence of each taxon is independent of that of the others, given the water quality class, appears to hold for the vast majority of taxa. Thus the 'naive' Bayes model appears well-suited to this domain.

Two 'naive' Bayes models have been tested, differing only in the way in which their prior and conditional probabilities are derived. The first used a well-founded statistical method, the *m-estimate* (Cestnik, 1990), to estimate the 'true' conditional probabilities from their raw values. In his case:

$$P'(H_i|e_{jk}) = \frac{n_{ijk} + mP(H_i)}{N_{jk} + m} \quad (1)$$

where $P'(H_i|e_{jk}) =$ the revised conditional probability that the water quality H is class i, given evidence e_{jk} that taxon j is in state k.

n_{ijk} = number of occurrences in the data of taxon j in state k in class i ;

N_{jk} = total number of occurrences of taxon j in state k across all classes;

$P(H_i) =$ the prior probability of class i derived from the data set; and

m = a parameter of the estimation method.

The second was strictly based on the Principle of Indifference and used a simple *ad hoc* procedure to eliminate zeros from the raw conditional probabilities, by uniformly redistributing a small percentage of the raw distributions. In this case:

$$P'(H_i|e_{jk}) = (1-x)P(H_i|e_{jk}) + \frac{x}{M} \quad (2)$$

where: $x =$ fraction of total probability redistributed;

$M =$ number of classes;

$P(H_i|e_{jk}) = P(e_{jk}|H_i)/\Psi$, where $\Psi =$ the normalisation constant for $i = 1$ *to* M;

(NB. This follows from the application of the Principle of Indifference and the exhaustive nature of the classes $i = 1$ *to* M)

$P(e_{jk}|H_i) = n_{ijk}/N_i$, where $N_i =$ number of occurrences of class i in the data.

The significance of the parameters m and x is briefly discussed later Five different models of each type were tested: m = 0, 1, 2, 5 and 10; and x = 0, 0.001, 0.05, 0.10 and 0.15.

Neural Nets

Two types of neural network were tested: multi-layered perceptrons (MLP's) which use supervised learning based on the back propagation of errors; and self organised maps (SOM's) which use unsupervised learning based on Kohonen's feature mapping algorithm. The interested reader is referred to the text by Beale and Jackson (1990), which provides a good introduction to neural computing and includes details of the structure and training algorithms of the two networks used here.

The MLP's had 36 input nodes (one for each taxon), one output node and either one, two or three hidden layers. Three networks had just one hidden layer (MLP60, MLP80 and MLP120) with six, eight, and twelve nodes respectively. Two had two hidden layers: MLP63 with six nodes in the first layer and three in the second, and MLP84 with eight in the first and four in the second. The hyperbolic tangent transfer function was used in all cases and training was achieved using the Quick Propagation algorithm. The inputs were not re-scaled, thus their range remained -0.2 to +0.9. The target values for the single output node in the training mode were the CI values of the samples, which ranged from 1 to 5, whilst the predicted CI's during testing were found to range from 0.86 to 5.17.

The SOM's were single layer networks in which the 36 input nodes mapped onto a rectangular feature map. These networks learn by recognising different patterns in the data and allocating each, in terms of its feature vector, to a specific output node in the feature map. This is achieved in a way such that adjacent nodes represent similar patterns, thus producing a matrix of gradually varying feature vectors, hence the term feature map. Since training is achieved without any knowledge of thc samples' classifications, the pattern represented by each node has to be identified and labelled to give it meaning. This was easily achieved in this study by examining the expert's prior classifications of the samples allocated to each node. Once a network is trained and labelled, it classifies samples by allocating them to a node in the feature map on the basis of the best match between their features and the node's feature vector. For the purpose of this study, three nets were trained: SOM1x10, SOM3x8 and SOM5x5 having 1x10, 3x8 and 5x5 feature maps respectively.

Regression and Model Trees

Regression trees (Breiman *et al.*, 1984) represent a functional dependency between a continuous valued class and several independent variables, which is not uniform across the whole domain. The leaves of the tree correspond to sub-domains where the dependency can be approximated to a constant (regression tree) or linear model (model tree). The nodes of the tree correspond to independent variables and the branches to tests on these variables. The M5 system for the induction of model trees (Quinlan, 1993) was used to build the model tree (MT1) shown in Figure 1 using the 36 attributes of the data set (i.e. the indicator taxa). To understand how the tree should be interpreted, consider a sample in which Heptageniidae is present, Leuctridae is absent and Nemouridae is few (i.e. the corresponding attribute has a value of 0.2). Moving through the tree, we end up at the leaf in which the Classification Index, CI = 4.26 + 1.53 Nemouridae, and obtain the prediction that CI = 4.26 + 1.53 x 0.2 = 4.87.

Since MT1 was found to be not very accurate, giving only 58.8% correct classifications, M5 was provided with six additional attributes that describe the diversity of the community. The result was the model tree (MT2) depicted in Figure 2. This tree is smaller but more accurate than MT1. Only one of the six attributes, the number of taxa present, was found to be relevant. In fact, it turned out to be the most relevant attribute.

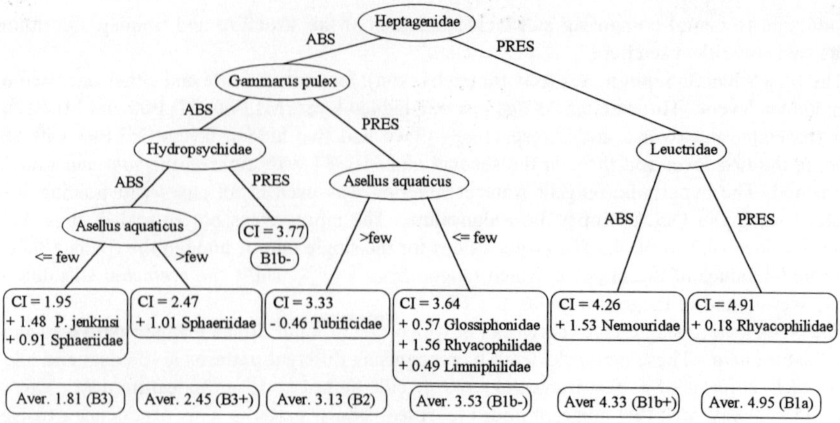

Figure 1 **Model Tree No.1 (MT1), derived from 36 attributes.**

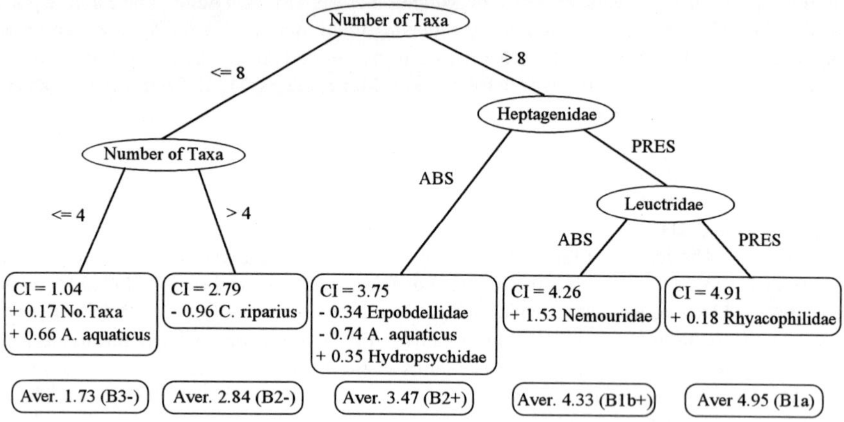

Figure 2. Model Tree No.2 (MT2), derived from 42 attributes.

4 RESULTS OF PERFORMANCE TESTS

Two different performance measures were used to assess the performance of the various models: their classification rates and uniformity of precision across the classes. These were further divided into two types of precision: a) the percentage of correct categorical classifications into classes B1a, B1b, B2, B3 and B4; b) the percentage of predicted CI values lying within 0.25 (narrow-band criterion) and 0.50 (broad-band criterion) of a class interval. Since the outputs of most of the models were in terms of the CI scale, the boundaries between the discrete classes were taken to lie at the 0.5 points on the scale. This assumption and that underlying the second set of tests strictly requires that the CI scale be an interval scale. That is, any given fraction of a class interval should mean the same in terms of change in water quality anywhere on the scale. However, since the scale was only used to compare like with like and not as an absolute interval scale, this issue clearly had little bearing on the value of the results as a means of comparing the relative performances of the models. Thus it was disregarded.

All models except the Bayesian models produced outputs on the CI scale. The Bayesian outputs were in terms of five probabilities, one for each of the discrete classes. These were converted to a single value on the CI scale by deriving the mean value, from:

$$\mathrm{CI} = \sum_{i=1}^{i=5} iP(H_i) \qquad (3)$$

where i is the class number (i.e. 1 for B4 through to 5 for B1a) and $P(H_i)$ is the predicted probability of class i. The overall classification rates of the best performing models in each category are given in Table 1.

Table 1. Overall classification rates for selected models

	Percent. Correct Categ.	*Predicted CI's within stated fraction of a class interval (%)* <0.25	<0.50
ASPT/1.28	76.3	49.5	79.4
Bayes (x =0.10)	84.5	71.1	88.7
Bayes (m=1)	83.5	67.0	85.6
Bayes (m=5)	85.6	62.9	86.6
MT2	73.2	53.6	79.4
MLP63	82.5	61.9	88.7
MLP84	81.4	66.0	90.7
SOM3x8	74.2	50.5	77.3
CM84(x=0.1)	82.5	69.1	92.8
CM84(m=5)	83.5	66.0	93.8

Once these figures were known it was decided to combine the best two models on a 50/50 basis to form a 'consensus' model. In fact, two consensus models were formed: CM(84/x=0.1) which combined MLP84 with Bayes(x=0.1); and CM(84/m=5) which combined MLP84 with Bayes(m=5). The classification rates achieved by these are shown at the bottom of Table 1.

Table 2 shows the distribution of correct categorical classifications across the five classes for each of the models listed in Table 1. The standard deviations of the success rates are provided as a measure of the models' variability in performance across the classes. Table 3 shows the equivalent statistics for the broad-band and narrow-band precision tests. The figures under each class heading represent the percentage of samples in the class (i.e. incl. +ve and -ve subclasses) with predicted CI values lying within the stated range of the expert's classification.

Table 2. Distribution of percentage correct categorical classifications across the five quality classes for selected models

	B1a	*B1b*	*B2*	*B3*	*B4*	*Stnd Dev.*
ASPT/1.28	68.4	81.8	78.1	66.7	88.9	8.3
Bayes(0.1)	89.5	68.2	93.8	86.7	77.8	9.1
Bayes(m=1)	89.5	63.6	96.9	86.7	66.7	13.1
Bayes(m=5)	84.2	68.2	96.9	86.7	88.9	9.4
MT2	57.9	77.3	81.3	80.0	55.6	11.3
MLP63	100.0	68.2	90.6	80.0	55.6	15.8
MLP84	100.0	63.6	90.6	80.0	55.6	16.5
SOM3x8	84.2	50.0	90.6	53.3	88.9	17.9
CM(84/x=0.1)	89.5	59.1	96.9	86.7	66.7	14.4
CM(84/m=5)	94.7	59.1	96.9	86.7	66.7	15.2
Average	85.3	66.4	91.3	78.0	71.2	

Table 3. Distribution of percentage classifications correct to within <0.25 and <0.50 of a class interval for selected models

	<0.25 of a class interval					*Stnd*	*<0.50 of a class interval*					*Stnd*
	B1a	*B1b*	*B2*	*B3*	*B4*	*Dev.*	*B1a*	*B1b*	*B2*	*B3*	*B4*	*Dev.*
ASPT/1.28	47.4	54.6	46.9	33.3	77.8	14.6	73.7	90.9	81.3	60.0	88.9	11.3
Bayes(x=0.1)	79.0	63.6	84.4	46.7	66.7	13.1	89.5	86.4	96.9	80.0	77.8	6.8
Bayes(m=1)	68.4	45.5	87.5	53.3	66.7	14.4	89.5	77.3	96.9	73.3	77.8	8.8
Bayes(m=5)	68.4	40.9	81.3	46.7	66.7	14.9	84.2	77.3	100.0	73.3	88.9	9.3
MT2	57.9	63.6	56.3	53.3	11.1	19.0	63.2	81.8	84.4	86.7	77.8	8.3
MLP63	79.0	54.6	68.8	40.0	55.6	13.3	100.0	77.3	93.8	86.7	77.8	8.9
MLP84	84.2	59.1	75.0	33.3	66.7	17.3	100.0	77.3	93.8	93.3	88.9	7.6
SOM3x8	78.9	31.8	53.1	40.0	44.0	16.2	84.2	68.2	87.5	53.3	88.9	13.7
CM(84/x=0.1)	84.2	59.1	78.1	46.7	66.7	13.4	89.5	86.4	96.9	93.3	100.0	4.9
CM(84/m=5)	79.0	54.6	78.1	40.0	66.7	14.8	94.7	81.8	100.0	93.3	100.0	6.7
Average	72.6	52.7	70.9	43.5	58.8		86.9	80.5	93.1	79.3	86.7	

5 DISCUSSION

River ecologists are unquestionably the real experts in the field, well able to perform complex mental mappings from data to water quality, but the data interpretation models which they have devised to date are too simplistic to produce adequate levels of performance. Two of the AI techniques tested in this study, 'naive' Bayes and multi-layered perceptrons, significantly out-performed the two traditional methods tested. However, before appraising their performances in detail it is worth considering two domain specific factors.

Domain Specific Considerations

Firstly, categorical classes and continuous indices are both in common use. River quality is commonly defined in terms of discrete classes, normally five as in this study, and these are displayed in different colours on river maps. This is done primarily to facilitate the dissemination of information to the public and politicians in an easily understood form. In an operational setting, however, it is often necessary to be more specific, so that one can determine how quality is changing in time or space, or whether a site is near the top or bottom of its stated class. For these purposes more precise continuous indices of quality are required.

Secondly, the frequency of the quality classes is not uniformly distributed. There are fewer B3 and B4 rivers than there are B1a and B1b, but precision in classification is just as important in poor quality rivers as it is in good ones, perhaps even more so. Consequently, a good model must not only achieve a high overall proportion of correct classifications, but this precision must be fairly uniformly distributed across the classes. Thus the variability of the success rates across the classes, as reflected by their standard deviation, should be as small as possible.

Overall Performance Rankings

In order to achieve an overall measure of relative performance, each model was ranked with respect to its success rate and uniformity of distribution using the results given in Tables 1, 2 and 3. The score, on which the overall ranking was based, was calculated as the weighted sum of the rankings on each test using a weight of two for the classification tests and one for the uniformity tests. It is clear from the results of this analysis (Table 4) that Bayesian inference was the most successful of the AI techniques tested, followed by multi-layered perceptrons.

Table 4. Rankings of models based on classification rates and uniformity of precision

	Ranking on Classif'n Rates				*Rank'g on Unif. of Precision*				*Overall*	
	Categ.	*<0.25*	*<0.5*	*O'all*	*Categ.*	*<0.25*	*<0.5*	*O'all*	*Score*	*Rank'g*
Bayes(x =0.1)	2	1	4	1	2	1	3	1	20	1
CM(84/x=0.1)	6	2	2	3	6	3	1	2	30	2
CM(84/m=5)	3	5	1	2	7	6	2	5	33	3
Bayes(m=1)	4	3	7	6	5	4	6	4	43	4
Bayes(m=5)	1	6	6	4	3	7	8	7	44	5
MLP84	7	4	3	5	9	9	4	9	50	6
MLP63	5	7	5	7	8	2	7	6	51	7
ASPT/1.28	8	10	8	8	1	5	9	3	67	8
MT2	10	8	9	9	4	10	5	8	73	9
SOM3x8	9	9	10	10	10	8	10	10	84	10

The two consensus models CM(84/x=0.1) and CM(84/m=5) were beaten only by the Bayes(x=0.1) model, which was the overall best performer on both the classification and the uniformity rankings. However, the CM's did produce the highest success rates on the broad-band (<0.5) criterion, but there was little evidence that a consensus approach is capable of producing a significant improvement relative to stand alone models. The poor performances of the SOM's and the TBI, which was eliminated at an early stage, were partly due to them being categorical classifiers not well suited to use on continuous scales, but even so they did not perform particularly well as categorical classifiers.

The difficulties which most models had in achieving high performance levels on classes B1a and B3 were largely due to the fact that these two classes had high proportions of samples with + and - classifications. Thus, they had a high percentage of borderline cases and presented a greater challenge to the models.

Traditional Methods

In order to offset the possibility of ASPT and TBI being disadvantaged due to them not having been 'trained' on the training set, they were given the benefit of being calibrated to the CI scale by linear regressions on the test set (section 3). In addition, the expert who classified the samples in the database was a leading figure in the development of the system on which the ASPT is based. In fact, he chaired the working sub-group of the committee which produced the system, so his thinking is 'in tune' with ASPT. Thus the authors believe that, if anything, ASPT was advantaged and not disadvantaged in these tests.

Bayesian Models

Much of the success of the Bayesian models was due to their assumption that the data was unreliable, especially with respect to the total absence of some taxa from some classes. Both of the methods tested, m-estimate and *ad hoc* zero-elimination, effectively eliminated zero probability values from all classes. The impact which this had on performance can be seen from the results of the x=0 (i.e. zeros remain) and x=0.1 (i.e. zeros replaced by 0.02) models, which gave 70.1% and 84.5% correct categorical classification respectively. The effect of the different assumptions underlying these two methods is difficult to discern in the results. Both models performed very well, although Bayes(x=0.1) out-performed Bayes(m=1) and Bayes(m=5) on both classification rates and uniformity of precision. Whether or not this was statistically significant is difficult to say. Clearly, the different assumptions about prior probabilities, $P(H_i)$, must have produced different distributions of precision, thus the application of the Principle of Indifference in Bayes(x=0.1) may have been responsible for its superior uniformity figures. However, this requires further investigation before firm conclusions can be drawn.

Multi-layered Perceptrons

The multi-layered perceptrons did not achieve quite the same levels of performance as the Bayesian models, but they have scope for improvement, both in terms of their internal formulation and the coding their data inputs. Furthermore, they will readily facilitate more complex formulations of the water quality mapping, involving environmental factors, simply by the addition of the appropriate variables to the input vector. Thus, they are considered strong contenders for use in future systems, either acting alone or in conjunction with Bayesian models. The Bayesian and neural network approaches are seen as complementary and thus well-suited for use in systems which take a consensus view.

Model Trees

Although the model tree (MT2) finished below ASPT overall, this was only due to its exceptionally poor performance (11.1%) on just one class (B4) in the narrow-band tests. This resulted in it finishing last in the narrow-band uniformity test, despite achieving excellent uniformity over classes B1a to B3. Although their performance was relatively poor, it is worth noting that the model trees used very few of the 36 indicator taxa (see Figures 1 and 2), a fact which could be of practical value. MT1 used twelve of the taxa, five in the decision tree and seven in the regression equations. MT2, which performed much better than MT1 (e.g. 73.2% to 58.8% on categorical classifications), used only eight of the taxa, two in the decision tree and six in the regression equations. The key to MT2's better performance was its use of the 'Number of Taxa' in the decision tree and the regression equations. In effect, this pooled some of the information from the missing twenty-eight taxa into one statistic, but too much was lost to enable the model to match the Bayesian and MLP models. They had the benefit of using the information provided by all 36 taxa. Nevertheless, the model trees do have one advantage over the other models in that they are transparent. Their reasoning is clearly apparent to the user, and in the case of MT1 and MT2 this reasoning has been confirmed by H. A. Hawkes as being broadly consistent with expert knowledge, albeit somewhat limited in its scope. Given a larger training set, these models may be able to achieve greater sophistication and hence better performance, but at present the benefit of their transparency is insufficient to compensate for their poor performance.

6 CONCLUSION

A variety of computer-based water quality classification models have been developed and tested alongside two traditional models, namely ASPT and TBI. Three different AI techniques were used to develop the computer-based models: 'naive' Bayesian inference, neural networks and machine learning (model trees), and two different formulations of the Bayesian and neural models were used.

The overall best performer was judged to be the Bayes(x=0.1) model, but this was closely followed by two systems CM84(x=0.1) and CM84(m=5) which took the consensus view of a Bayesian model and a multi-layered perceptron. Taken overall, two types of system clearly out-performed ASPT (the best of the traditional models): the Bayesian models and the multi-layered perceptrons, but the former marginally out-performed the latter. However, in view of the scope for further improvement to the multi-layered perceptrons it seems that they, together with the Bayesian models, are the most promising techniques for use in future bio-monitoring systems, either acting alone or together as complementary 'experts'. A cross-validation evaluation of the methods is necessary in order to provide a more reliable estimate of their relative performances.

The key to the realisation of the full potential of bio-monitoring lies in the development of reliable data interpretation and classification systems. Biological data are potentially very rich in information, but the mappings necessary to realise this are complex. This paper has demonstrated the potential of some AI techniques to provide such mappings and has provided pointers to future developments.

7 ACKNOWLEDGEMENTS

The authors wish to thank H. A. Hawkes for his invaluable assistance in classifying the biological samples, and also to the NRA (Severn-Trent Region) for providing the data. Thanks are also due to Ross Quinlan for providing a copy of his M5 system, and to Brendan Ruck for constructing the database.

8 REFERENCES

Beale R. and Jackson T. (1990) *Neural Computation: An Introduction.* Adam Hilger, Bristol and New York.

Breiman L., Freidman J.H., Olshen R.A. and Stone C.J. (1984) *Classification and Regression Trees.* Wadsworth, Belmont.

Cestnik B. (1990) Estimating Probabilities: A crucial task in machine learning, in *Proc. European Conference on Artificial Intelligence.* Stockholm, Sweden.

De Pauw N. and Hawkes H.A. (1993) Biological monitoring of river water quality, in *River Water quality Monitoring and Control,* ed. W. J. Walley and S. Judd. Aston University, Birmingham. 87-111.

Kolkwitz R. and Marsson M. (1902) Grundsätze für die biologische Beurteilung des Wassers nach seine Flora und Fauna. *Mitt. Prüfungsanst. Wasserversung. Abwasserbeseit.* **1**. 33-72.

Michie D., Spiegelhalter D.J. and Taylor C.C. (Eds) (1994) *Machine Learning, Neural and Statistical Classification.* Ellis Horwood.

Quinlan J. (1993) Combining instance-based and model-based learning, in *Tenth International Conference on Machine Learning.* Morgan Kaufman. 236-243.

Ruck B.M., Walley W.J. and Hawkes H.A. (1993) Biological classification of river water quality using neural networks, in *Applications of Artificial Intelligence in Engineering VIII,* eds. Rzevski G., Pastor J. and Adey R.A. Elsevier/CMP. 361-372.

Walley W.J., Hawkes H.A. and Boyd M. (1992) Application of Bayesian inference to river water quality surveillance, in *Applications of Artificial Intelligence in Engineering VII,* eds. Grierson D.E., Rzevski G. and Adey R.A. Elsevier/CMP. 1030-1047.

6 BIOGRAPHY

William Walley was formerly Senior Tutor in Civil Engineering at Aston University, where he taught water resources systems and researched in the field of computer modelling. In 1989 he teamed up with ecologist H. A. Hawkes to develop AI methods of river quality classification. He moved to Staffordshire University's School of Computing in 1993 to consolidate this work.

Sašo Džeroski is a researcher at the Jožef Stefan Institute, Ljubljana, having previously been a visiting researcher at the Turing Institute, Glasgow, and the Computer Science Department of the Katholieke University Leuvens, Belgium. He has researched into various areas of the theory and application of machine learning. He gained his BSc, MSc and PhD degrees from the Faculty of Electrical Engineering and Computer Science of Ljubljana University.

21

On incorporating physical building parameters into neural net models of dynamic thermal behaviour

N. ROBINSON, S. BUREK*, G. BURNS#*
**Department of Energy and Environmental Technology, Glasgow Caledonian University,*
Cowcaddens Road, Glasgow, G4 OBA, Scotland.
e-mail nro@uk.ac.gcal

#Department of Engineering, Glasgow Caledonian University, Cowcaddens Road, Glasgow, G4 OBA, Scotland.

Abstract
This paper investigates the ability of feedforward artificial neural networks to distinguish between data-sets from a dynamic model of the thermal behaviour of buildings with different structural and operational parameters. Results are presented of preliminary attempts to incorporate both physical building parameters and dynamic data into a general model. Once developed, this technique should enable use of data-based ANN models as mechanistic investigative tools.

Keywords
Neural networks, building, thermal behaviour, modelling

1 INTRODUCTION

1.1 Context

Fully-connected feed forward neural networks are well suited to modelling complex or only partially understood systems for several reasons - for example: due to bounded transfer functions, they can mimic highly non-linear sequences (Montague and Morris, 1994) and they are capable of adopting a wide variety of mathematical structures due to their ability to emphasise or ignore inputs. The ability to predict time-series data in the absence of a mechanistic understanding of the system is clearly useful for practical control purposes. Through learning from examples, an artificial neural network (ANN) stores information, describing the system, in its weights and biases.

Models of building thermal behaviour can be used to design energy-efficient buildings or assess the likely benefits of energy-saving measures. Many such models exist (Clark and Irving, 1988) employing a wide variety of modelling techniques featuring analogues of what the modeller considers to be the dominant physical mechanisms at work. An artificial neural network model requires fewer assumptions regarding its mathematical structure. However, it can only be expected to accurately predict the behaviour of the building which produced its training data.

The authors are attempting to develop a method of extracting meaningful physical building information from data-based ANN models of dynamic building behaviour by training a general model on a mixture of dynamic and pre-determined parameters defining the structure of the building and its mode of operation. Once developed, such a technique could be very easily translated into other areas of environmental or policy-making software design for use in areas where there is a database but limited understanding and in which it may be helpful to be able to make predictions concerning the outcome of a set of actions or policies.

Before using data from real buildings, it was decided to develop the technique using computer-generated data. A computer model of a greenhouse was used to generate several sets of hourly temperatures and auxiliary heat inputs as training data for different greenhouse structures and modes of operation.

1.2 Rationale

ANN models trained to be specific to each data-set were considered useful in establishing the nature of the data to be used in this study. For the ANN models to be considered specific to the data on which they were trained, they should yield significantly higher prediction errors when applied to the other data-sets. In the case that they are shown to be specific in this way, they must contain information representing the relative values of the greenhouse parameters used in generating the data-sets.

The degree to which each model is specific to its training data may be considered to be a measure of the relative importance of the parameter set (structural and operational information) which distinguish it from the other greenhouses.

A general model, capable of predicting the heat input, Q, to the same accuracy as a specific model, would need to alter its mode of operation according to the structural and/or operational greenhouse parameters used to generate the data, Therefore, it should need access to this information as additional inputs. Such a general model would allow the investigation of the links between a building's architecture and its thermal performance.

2 METHOD

Each data set was modelled using its own ANN model. All were feedforward neural networks with sigmoidal transfer functions in the hidden layer neurons and linear transfer functions in the output neurons. The ANNs were trained using Jacobs' (1988) delta-bar-delta back-propagation algorithm with a momentum term. Feedforward neural networks were chosen in preference to recurrent neural networks because although recurrent neural networks, are well-suited to modelling dynamic behaviour (Si and Pang, 1993), they are inherently more difficult to dissect. A feedforward neural network can be used to model dynamic time-series by carrying out static mappings using past outputs as inputs in addition to the suspected system inputs.

The ANNs were trained and analysed on a PC and coded using C. For an introduction to artificial neural networks, see, for example Haykin (1994).

2.1 Creation of specific ANN models

A complex simulation of the thermal behaviour of greenhouses, based on fundamental heat transfer principles, was used to generate hourly data for the heating and internal temperature, given ambient weather data. Twenty-five sets of hourly data were generated, each set representing the behaviour of a greenhouse with slightly different physical or operating parameters to the rest. Each set contained one year of hourly internal temperature and heat input information along with weather information.

All the data was scaled from minus one to plus one. The weather data was common to all the data sets but the range of the heating input(Q) and the internal temperature(T) varied from one data set to another. In order for it to be possible for all the data sets to be modelled by one general model, a common scaling had to be used for all data sets. In order to ensure that data sets with large ranges did not dominate the modelling process and the subsequent analysis, the RMS error derived from attempting to map samples from any one data set is reported as a percentage of the range of Q for that data set.

Of the 8760 sequential input-output pairs in each data set, 1600 pairs were selected at random to be used as a training set and the rest as a validation set. The training sets were divided in half. 800 samples were used to adjust the weights and biases of the network. The other 800 pairs were used as a sample test set, to gauge to what extent the model would generalise to data which had no direct influence on its training.

Twenty-five ANN models (one for each data set) were trained, using back propagation, to predict a value for the heat input (Q) at time t, based on the values of Q and T (the internal temperature) at time t-1 hours and the weather data (solar radiation and ambient temperature) at time t.

Each of the twenty-five ANN models consisted of six neurons in the input layer - four for the hourly weather data and one for each of the delayed outputs (Q(t-1) and T(t-1)). There were two neurons in one hidden layer and one neuron as the output Q(t). The number of neurons in the hidden layer was chosen using an ANN feature detector as described by Montague and Morris (1994).

The criterion for the training was the output, Q(t) from the ANN, compared with the value for Q(t) generated by the mechanistic computer simulation and the difference termed the error. The set of weights which yielded the smallest root-mean-squared (RMS) error when applied to the sample test set was saved and training continued until no significant improvements were being made. Each ANN model was then validated using its own validation set and found to produce a similar RMS error to that obtained on the sample test set. The extent to which these models were specific to their own data was explored by running each model with every validation set and examining their RMS errors (figure 1).

2.2 Creation of general model

A general ANN model was trained using all twenty-five sets of training data. In addition to the six input fields mentioned above, another seven were added to allow input of parameter values. It was anticipated that this model would use the parameter data to make appropriate predictions.

To act as a control, another ANN model was trained on all twenty-five sets of training data but was not given access to parameter data.

The general model and the control were deliberately over-specified - both ANN models contained two hidden layers with thirteen neurons in the first and ten in the second. Overfitting was avoided by using a sample test set and saving the best nets as they arose.

2.3 Summary of terms

- **Specific models** are each trained using weather data, internal temperature and heating data only. That is to say that there is no explicit input of structural or operating parameters though these are implied by the dynamic data.
- **Intended data set** refers to the validation data set that a specific ANN model was trained to be able to map.
- The **general model** is trained with the weather, internal temperature and heating data used to train all twenty-five specific models plus the structural and operating parameters used to create all twenty-five data sets.
- The **control model** is trained in the same way as the general model but is not given the structural and operating parameters as inputs.
- Using a standard shorthand, the specific ANN models had a 6-2-1 architecture (6 inputs, 2 hidden layer neurons, 1 output) whilst the general model can be described as 13-13-10-1 and the control model as having a 6-13-10-1 architecture (i.e. the control model is missing the seven parameter data inputs).

3 RESULTS AND DISCUSSION

All twenty-five models were successfully trained to 3 ± 0.75%. The results of running each model with each validation data-set are shown in figure 1. Figure 1 shows a steep sided valley running diagonally through the matrix of prediction errors. The points on this diagonal represent those runs in which the specific models were applied to their appropriate validation sets. The predominance of relatively high errors to both sides of this diagonal indicate there is a significant degree of specificity to the models. The presence of low error points away from this diagonal indicates that several models with different operating or structural parameters can share similar modes of behaviour.

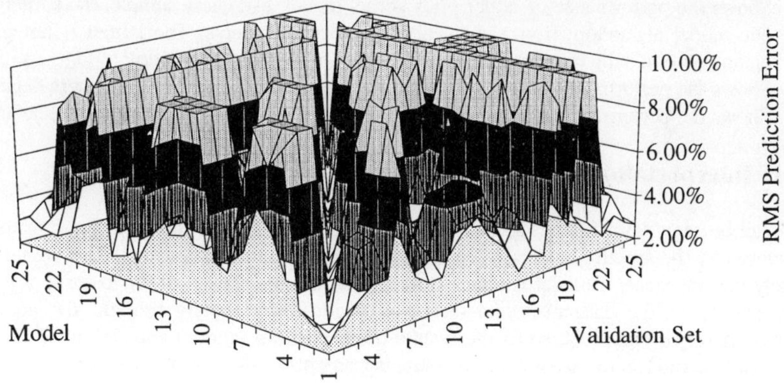

Figure 1 prediction error produced by running all models with all data sets.

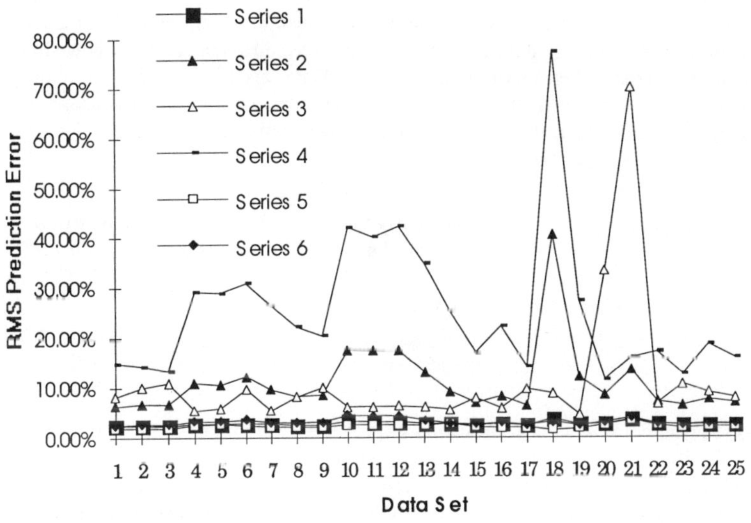

Figure 2a RMS prediction error plotted against data-set. Series:

1. Shows prediction error against validation set for the specific models run on their intended data sets.
2. Shows the average RMS error across all twenty-four ANN models when applied to each validation set.
3. Gives an error value for each model when tested across all validation data sets (in this case the x-axis represents the twenty-five specific models rather than their validation data sets).

4. Shows the performance of rather poor kernel model. For every sample, the kernel chooses one model at random from the twenty-five specific models. The kernel is not given the parameter data which might enable it to choose an appropriate model.
5. Shows the performance of the general model (see figure 2b, series 2, for more detail).
6. Shows the performance of the control model (see figure 2b, series 3, for more detail).

3.1 Interpretation and analysis

As can be seen from figure 2a, the performance of the general model at least matched that achieved by the specific models when tested on their own validation sets. This would seem to imply that the general model is able to use the parameter data in order to make a prediction appropriate to the particular data-set being used. Unfortunately though, the control also performs very well compared to the results obtained from random application of the twenty-five specific models to the problem of mapping any particular sample (i.e. figure 2a, series 4). Figure 2a series 2 and series 3 illustrate the extent to which the specific models are specific to their own data.

Examination of figure 2a, series 3, reveals that models 20 and 21 yield distinctly high average RMS error values when tested across all data sets. This implies that these models are more specific to their own training data sets than the other models. Also from figure 2a, series 2, data set 18 yields a high average RMS error value when used in testing across all models. This implies that data set 18 contains data describing a distinctly different mode of behaviour (due to a significantly different set of operating parameters). Data sets 18, 20 and 21 were produced using parameter sets in which the set-point temperature of the greenhouse (i.e. the temperature below which the heating system operates) was varied as in table 1.

Table 1

Model	Set-point °C	Total Yearly Q (KJ/m^2)	Maximum hourly Q (KJ/m^2)
18	0	49 734	562
19	5	729 306	1350
20	15	6166 469	2704
21	20	10 594 300	2704

From a mechanistic point of view, one would expect large variations in set-point temperature to produce significant changes in the dynamic behaviour of the system as well as the total yearly energy used (i.e. at an average outdoor temperature of 11.25 °C, greenhouses 18 and 19 would experience a net gain in heat energy over the year whereas greenhouses 20 and 21 would experience the opposite).

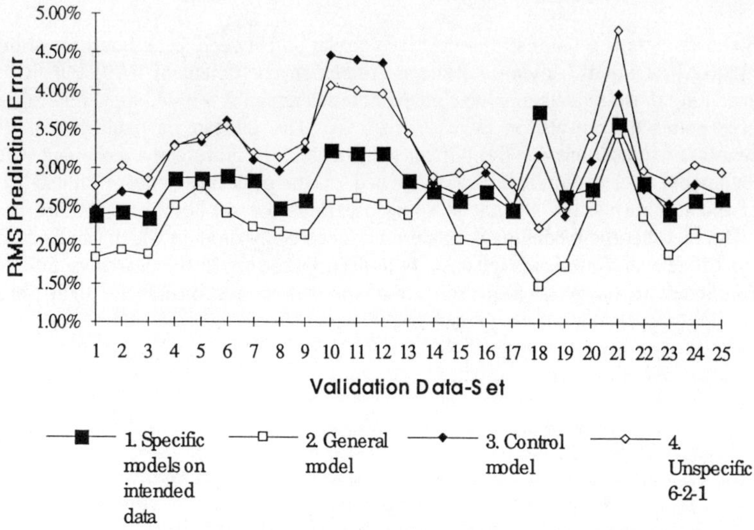

Figure 2b RMS prediction error plotted against validation data set

Figure 2b allows more detailed inspection of the relative performances of the general and control models tested across all validation data sets; the specific models when tested on their intended validation data sets; an unspecific 6-2-1 trained and tested across all data sets. From figure 2b, it can be seen that, in most cases, the control model performed less well than the specific models and that the general model performed somewhat better.

In the case of data set 18, the specific model performed worst of all. Repeated training of model 18 yielded no improvements even though the training and sample test set errors could be could be as little as half of that found when testing on the validation set. Further attempts using a 6-1-1 architecture yielded similar results seeming to imply that data set 18 contains 'simple' data and that the use of two or even one hidden layer neuron leads to an over-dimensioned[1] network. This is borne out by the relatively good performance of an unspecific 6-2-1 network when tested on data set 18. It seems reasonable to suggest that in the case of the general model and that of the unspecific 6-2-1, data set 18 formed part of a larger pattern. Whereas in the case of model 18, faced with simple data and an excess of input information and processing capacity, it could not help but overlearn.

In the case of data set 21, it can be seen that the general and specific models failed to improve significantly on the performance of the control model. This implies that although the simple architecture used by the specific ANN models proved sufficient to model the output of the mechanistic greenhouse model for almost all parameter sets, it cannot match the mechanistic model's capacity for complex behaviour.

3.2 A mechanistic explanation

Figure 3 shows there is a fairly strong positive correlation between 'peak hourly heat input' and 'total yearly heat input' (giving a Pearson correlation coefficient of 0.79) but that the points representing those models in which 'set-point temperature' is varied, appear as rogues (without these points the correlation increases to 0.96). This indicates a likelihood of there being mechanistic explanations for the difficulty encountered in accurately modelling certain modes of behaviour (for example, in models 20 and 21, the peak heat input is limited by the capacity of the heating system so that at some occasions the set-point temperature is not achieved). The fact that the modelling procedure performs badly on data which can be seen as significantly different in a mechanistic sense, is in itself promising in the search to link data-based ANN models to the physical parameters of the systems responsible for their training data.

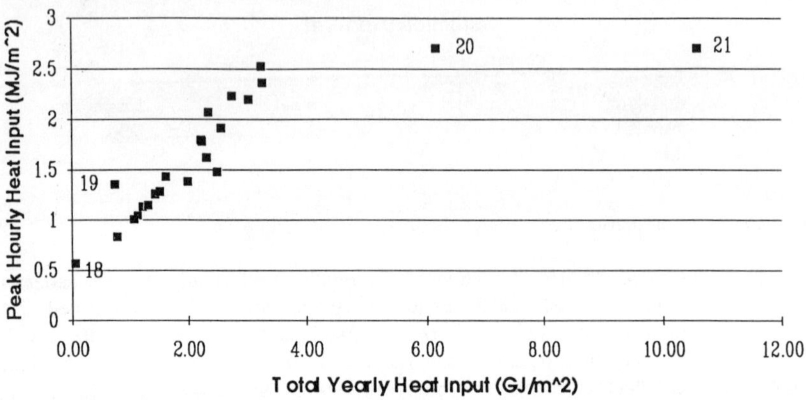

Figure 3 peak hourly heat input plotted against total yearly heat input.

ANNs with sigmoidal activation functions are recognised as 'universal approximators' (Mitchell, 1992) (i.e. they are able to map any continuous function) and therefore a sufficiently large and complex architecture, successfully trained, should be able to reproduce the mechanistic greenhouse model in all its complexity. The 'error-backpropagation' training algorithm, being a gradient-descent method, has a tendency to yield sub-optimal solutions (Willis et al., 1991; Marshall and Harrison, 1991). This has become a problem when trying to establish whether or not a particular ANN architecture is capable of matching the dynamic behaviour produced by any/all configurations of the mechanistic model. It is for this reason that the authors are currently investigating several network training algorithms which claim to be globally convergent (Willis et al., 1991; Marshall and Harrison, 1991; Yakowitz, 1993).

3.3 Further tests

In order to gauge to what extent the general model was able to make use of the parameter data, it was run with all twenty-five parameter sets in turn over all twenty-five validation data sets in turn. From this it was possible to build up a cross-testing matrix of six hundred and twenty-five RMS prediction errors, analogous to that shown in figure 1. The result of this test was a three-dimensional surface plot which, at a glance was the same as figure 1 (and so is not included here). This was taken to imply that the general model was indeed able to make use of the parameter data as, by counter-example, inappropriate parameter data caused the general model to make erroneous predictions. Further inspection of the two cross-testing matrices (used to create the 3-d surface plots) confirmed that the general model was at least as specific in matching parameter sets to validation data sets as the twenty-five specific models were to their intended validation data sets. The ability of general models as described here, to predict the behaviour of a new system when given its parameter data, is still under investigation.

4 CONCLUSIONS

The specific ANN models performed generally better on their own data sets (i.e. those from which their training sets were taken) than on the other data sets. This implies two things:-

1. the data-sets used in this study described dynamic systems which were noticeably different at the resolution at which this study was carried out.
2. the individual models had been trained to a level of accuracy which made them specific to the data-set from which their training data was taken.

Therefore, this implies that the specific ANNs must have at least partial knowledge of the parameter values used in creating the data - stored in the form of the weights and biases defining each ANN model.

For a single general model to perform as well as these specific models across all twenty-five data-sets, it would need to encompass a range of modes of operation, and to switch between them depending on which particular data-set it was being asked to imitate. The general model was trained with input information connecting sets of parameters to modes of behaviour and, once trained, the general model used this parameter information to select an appropriate mode of operation.

The results presented here were obtained using data from a mechanistic, well-behaved computer simulation of the thermal performance of a relatively simple greenhouse structure. These results appear promising and the authors are in the process of developing and applying the techniques outlined here to real thermal performance data from real buildings.

5 REFERENCES

Montague G. and Morris J. (1994) Neural-network contributions in biotechnology, *Trends in Biotechnology*. **Vol. 12**, 312.

Clarke J. A. and Irving A. D. - guest editors. Special Issue devoted to Building Energy Simulation. *Energy and Buildings.* **Vol. 10 No. 3**.

Jacobs R. A. (1988) Increased rates of convergence through learning rate adaptation. *Neural Networks* **1,** 295-307.

Si J. and Pang L.(1993) Recurrent neural networks for dynamic system modelling, *Proceedings of the 1993 International Symposium On Intelligent Control.* 364-369.

Haykin S. (1994) *Neural Networks: A Comprehensive Foundation.* Macmillan, New York.

Mitchell J. (1992) Comparing feedforward neural network models for time series prediction, *Neural Computing Research and Applications,* **Part One,** 175-182.

Willis M. J.; Di Massimo C.; Montague G. A.; Tham M. T. and Morris A. J. (1991) Artificial neural networks in process engineering, *IEE Proceedings D,* **Vol. 138 No. 3** 219-266.

Marshall S. J. and Harrison R. F. (1991) Optimization and training of feedforward neural networks by genetic algorithms, *Artificial Neural Networks - 2nd. International Conference,* 39-43.

Yakowitz S. (1993) A globally convergent stochastic approximation, *SIAM J. Control and Optimization,* **Vol. 31 No. 1,** 30-40.

6 BIOGRAPHY

Nick Robinson graduated with a BSc in environmental science and independent studies from Lancaster before moving to Glasgow Caledonian for an MSc in energy systems and environmental management. He is currently studying for a PhD at Glasgow Caledonian University.

Dr. Stas. Burek graduated from Cambridge in 1979 as an engineer and has worked in the energy field for the last fifteen years. He gained a PhD in solar energy from Cranfield in 1987 and is currently a Senior lecturer at Glasgow Caledonian University.

Dr. George Burns graduated from the University of Strathclyde with a Bsc in physics, then a PhD in electronics. He is currently a senior lecturer and divisional leader in the Dept. of Engineering at Glasgow Caledonian University.

22

An Intelligent Prediction System for Oil Spill Movement

Haojin Wang, Jan Wolter and Jungfu Tsao
Texas A&M University
Dept. of Computer Science College Station, TX 77843 USA.
Telephone: 409-845-1489. Fax:409-847-8578.
email: `wolter@cs.tamu.edu`

Abstract

The goal of the research reported here is to build an intelligent system for the prediction of oil slick movement to facilitate the emergency management in case of a marine oil spill. In this paper, three main aspects of this system are discussed. First, an overview of the system architecture presents the major components of the system and the interconnection between them. Then the methods for acquiring diverse types of knowledge and for fusing them into a functional system is described. Finally, issues related to the implementation of this system are discussed.

Keywords

oil spill movement, prediction, knowledge acquisition, knowledge integration

1 INTRODUCTION

1.1 Issues

The serious environmental, economic and social consequences of oil spill have been demonstrated by recent major oil spills. Previous research in the fields such as oceanography, marine-physics and others have concentrated on developing various mathematical models for oil slick movement on the surface of ocean. However, making precise predictions may never be possible, since neither the current state of the world nor future developments in the conditions such as weather, current and temperature on which an oil spill model depends, can be reliably known. Thus, it is not very meaningful or it is even misleading to give decision makers a single projection of oil slick movement. Instead, we seek to produce a set of projections of the most likely scenarios for the oil slick with the likelihood attached to each of the projections. This will give decision makers a more realistic picture of the real world. To achieve this, we must make sure that our estimate of how conditions could develop takes into account all of the available knowledge and information. However, to complicate the issue, much of the observed data is available only in an imprecise form. For example, the wind directions are only given in approximate terms such as "east," "northwest," or "northeast" while we are required to predict the trajectory of oil slick movement as precise as possible. Domain expert's knowledge usually can only be described by vague terms.

In addition to the uncertainty and impreciseness associated with data and operating environment, the diversity in types of knowledge and information needs to be taken into consideration. For example, there are forecasts generated by other experts, such as weather forecasts. Then, there are historical records that may have been collected for the spill region over the years about weather and currents. Also, the user of the prediction system serves as a domain expert to give the judgement on the on-going situation. Correspondingly, different knowledge acquisition methods are required.

The system architecture we developed in this research stresses the integration of a knowledge base with various mathematical models that simulate various oil motions in the ocean. At the core of the system is a knowledge acquisition module. The module features different methods for acquiring diverse types of knowledge and the fusion of different types of knowledge to produce a realistic assessment of the oil slick movement. An object-oriented paradigm is used to implement the system and major issues related to the implementation are discussed.

1.2 Related research

Intensive research activities on oil spill modeling first occurred in the 1970s as a response to potential danger the marine oil spill may pose and as a result of people's increasing awareness of importance of environmental conservation. The initial research was limited to certain disciplines such as oceanography, industrial engineering and others (Anderson, 1974; Fay, 1971; Williams, 1974). There are some characteristics common to all these research. They take input data as given and treat the issue of predicting oil slick movement as a purely mathematical modeling problem. The oil spill model, which model the oil slick movement, usually consists of several distint submodels. Each of the submodel simulates a type of oil slick motion, such as dispersion, weathering, spreading or transport. Weathering refers to reduction in amount of oil due to the factors such as evaporation, emulsification, and biodegradation of oil. This is affected by the type of oil and weather conditions (Samuels, 1982; Venkatesh, 1988). Transport refers to the motion of an oil parcel from one location to another and is mainly determined by surface wind and current. Dispersion refers to the process by which waves spread the oil across the water surface or into the water column. Spreading is the process in which the oil layer thins out due to gravity and surface tension (Venkatesh, 1990).

In a related project we have addressed the problem of interpreting recent observations of the weather and the oil movement (Tsao, 1993; Tsao, 1994). This in effect uses an oil spill model to construct a scenario that fits the available observations and agrees with our model's understanding of the behavior of oil slicks. This model-based understanding of uncertain observational data of the oil slick provides the initial conditions from which predictions of future conditions are projected. Since the model-based interpretation of oil slick behavior generally provides finer spatial resolution than forecasts, and since there is significant continuity in most parameters, this provides a good starting point for our predictions.

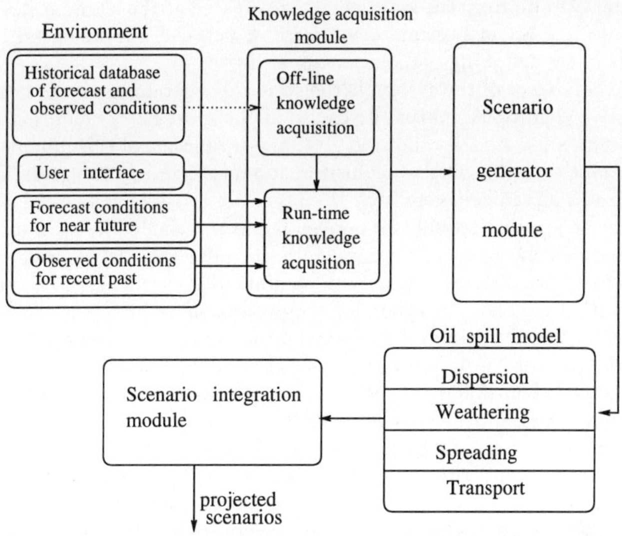

Figure 1 The Architecture for an Intelligent Prediction System

2 SYSTEM ARCHITECTURE

The intelligent prediction system consists of five distinct modules as depicted in Fig. 1. The functions each module performs and the connections between the modules are described in this section.

The first component of the system is the environment of the system which in turn consists of a user interface, a historical database of observed and forecast weather conditions, a set of forecast weather conditions for the near future and a set of observed conditions in the recent past. This environment module provides the sources from which the different types of knowledge are extracted during the learning stage as well as at run-time. The user interacts with the system through a user interface. There are two ways that a user can interact with the system. One is that the user enters the initial data values. For example, the user is prompted for the weather forecast, estimated volume of the initial spill, oil type, etc. The other way a user interacts with the system is that the user serves as a domain expert. The user is allowed to input the domain expertise into the system. One such expertise is his/her preference of one type of knowledge over another in deciding the most likely weather condition for the projection of the oil slick movement. The solid lines represent the run-time data/knowledge flow while the dotted lines the data flow during the learning stage.

The second component is the knowledge acquisition module. Its function is to build separate knowledge models based on the type of knowledge and the role each type of knowledge plays in generating a prediction and in inferring the likelihood of a given scenario. During the learning stage, the input to this module comes from the historical database of actual and forecast weather conditions. The run-time inputs come from users via a user interface and from the knowledge base created at the learning stage. The output of this module is a set of possible conditions. A detailed discussion on the knowledge acquisition is in the following section.

The third component of the system is the scenario generator module. It takes as inputs a set of possible conditions, makes the choice of the most likely condition and computes the likelihood measure for the condition. The major function of this module is to combine the results from the the knowledge acquisition module to satisfy two goals. One is to make a decision on the most likely condition from a set of possible conditions generated from the knowledge acquisition module for a given domain variable like wind or current. Then a set of the most likely conditions is input into the oil spill model to produce a projection of oil slick movement. The other goal is to combine the knowledge from different sources to come up with a likelihood measure for the projection.

The fourth component is an oil spill model that is adopted mostly from the existing oceanographic and marine physics research. This module consists of four distinct submodels that models four different types of oil slick motion, that is, weathering, transport, dispersion and spreading (Paluszkiewicz, 1989).

The last component, scenario integration module, takes as the inputs a set of projections of oil slick movement with the likelihood measures, and produces a set of projections of the most likely scenarios of the oil slick. The functions it performs include merging those projections whose differences are insignificant, deleting those that are highly unlikely, and preparing for the output to be presented to users.

Finally, the system graphically presents a set of the most likely scenarios of oil slick movement with their corresponding likelihood measure to the user who can use this valuable information to help make the oil spill management decisions.

3 KNOWLEDGE ACQUISITION (KA)

Diverse types of knowledge are involved in the reasoning process in this system. Some knowledge is acquired during the off-line learning period while other are acquired at run-time. Some are "precise" like statistical measures and some are less precise like domain expert's intuition or preference. Here we refer to the off-line statistical information extraction as a kind of learning. In the rest of this section, we focus on the methods for acquiring different types of knowledge.

3.1 Off-line knowledge acquisition

Different kinds of statistical data are acquired off-line during the learning period. Fig. 2 depicts the different submodules that acquire statistical data from different sources. Data and knowledge flow in the off-line KA is indicated in dotted lines while the run-time KA in solid lines.

We derived three different kinds of statistical data from historical database of actual and forecast weather conditions. The cycle analysis submodule is responsible for deriving daily and seasonal patterns of weather conditions. The statistical data is represented in the form of conditional probabilities, such as the following expression for daily pattern

$p_{ij}(Wind(t) \in W_j | t \in Hi)$

where H_i represents the ith hour of a day and W_j represents the jth wind (winds are

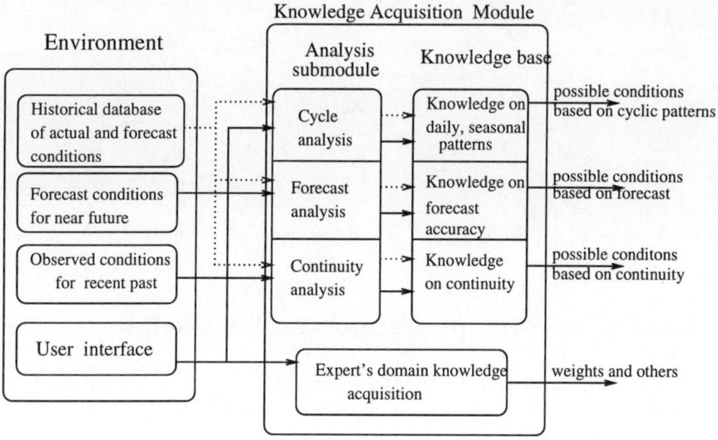

Figure 2 The Knowledge Acquisition Module

divided into m groups). p_{ij} is the conditional probability that the wind at H_i hour is W_j. Similarly, we can derive the weekly and seasonal patterns. The probabilities are easily obtained using standard statistic techniques.

The submodules "forecast analysis" and "continuity analysis" are responsible for deriving the knowledge on the forecast accuracy and the the continuity from the historical data. In the historical database, we have data on the past forecast and observed weather conditions and the difference between the two is the forecast error. The degree of continuity between two winds W_i and W_j are represented by the probability of transition from W_i to W_j. In a similar fashion, the knowledge on forecast accuracy and weather condition continuity is represented in probability tables.

These different kinds of statistical data are extracted off-line and stored in a knowledge base.

3.2 Run-time knowledge acquisition

In Fig. 2, the data and knowledge flow for the run-time knowledge acquisition is indicated by the solid lines. At run-time, when the current time and season are input into the knowledge acquisition module via a user interface, the cycle analysis submodule uses the daily, weekly or seasonal patterns that were built off-line to obtain a set of the most likely weather conditions based on the cyclic patterns. This is virtually a table look-up operation plus some constraint checking. In a similar manner, the forecast analysis and the continuity analysis submodules, using their respective statistical knowledge bases that have been built off-line, obtain a set of the most likely conditions based on their respective best knowledge.

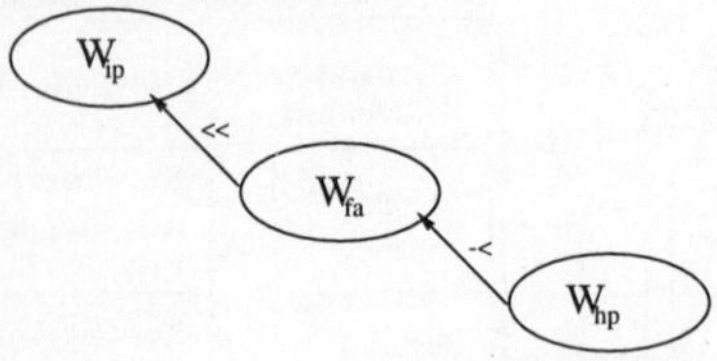

Figure 3 Influence Weight Graph for the Likelihood Measure of Weather Forecast

3.3 Acquisition of expert's domain knowledge

In Fig. 2, there is a link from the user interface to the submodule for acquiring expert's domain knowledge. The domain expert inputs his/her knowledge directly into the intelligent system via the user interface and the knowledge acquisition module extracts the domain knowledge and translates it into a form compatible with the statistical knowledge so that they can be processed together in the scenario generator module.

Heuristic knowledge such as domain expert's judgement and preference, and rules of thumbs like "near shore currents tend to run parallel to the shore" are often described in vague terms. For example, one of the major goals for this system is to generate a likelihood measure for a projection of oil slick movement. We identified three types of information which can influence the likelihood. They are the historical weather pattern for the season, the observed weather conditions of the recent past and the accuracy of the past weather forecast. However, the domain expert may favor one type of information over other, say, immediate past weather condition over historical weather pattern. The domain expert's preference and intuition are usually very difficult to quantify. We use a method described in (Weld, 1990) for extracting this type of imprecise knowledge. This method combines qualitative reasoning with fuzzy set theory (Zadeh, 1972) to acquire the influence weights. This method is based on the observation that domain experts usually hesitate to quantify the influence of variable X on Y. However, it is relatively easy for the domain experts to express the relative weights of X_i and X_j on Y. The method attributes fuzzy value to weights from the linguistic description of the relative orders of magnitude.

We define $X_1, ..., X_n$ be a set of variables influencing the variable Y with the respective weights $W_1, ..., W_n$. The linguistic description of orders of magnitude of the weights is a set of pairwise links between these weights and can be described by a formalism as shown below:

relation	verbal description
$A << B$	A is much smaller than B
$A- < B$	A is moderately smaller than B
$A == B$	A is approximately equal to B

If we use a node to represent a weight and an edge to represent the fuzzy link (linguistic descriptor) between the two nodes X_i and X_j, we have a directed acyclic graph as shown in Figure 3

In this example, the graph represents the following relations: $W_{fa} << W_{ip}$ and $W_{hp}- <$

W_{fa} where W_{fa}, W_{ip} and W_{hp} are the weights for the weather forecast accuracy, the immediate past weather condition and the historical weather pattern respectively. To compute the absolute value of the weights W_i, we assume that there exists a maximum weight W_{max} such that for any W_i, there exists a path from W_i to W_{max}. An algorithm to explore the weight graph from the top to the bottom in breadth-first fashion and to compute the intersect of the supports of each node's parent value along the way, will yield a set of numeric weights $(W_1...W_n)$ in $[0,1]$.

3.4 Integration of different types of knowledge

The different types of knowledge are fused to produce a projection of oil slick movement and a likelihood measure of the projection. The fusion takes place in the scenario generator module.

The domain expert's knowledge on the relative weight of each type of information is fused into the belief network to produce a likelihood measure for a projection. Given $(W_1, ..., W_n)$ in $[0,1]$, we can normalize the weights and then use the resulting weights to weigh the conditional probabilities:

$$\tilde{p}_i = p_i * \frac{W_i}{W_{max}}$$

where $\tilde{p}_i$ is weighted conditional probability and $W_{max} = max(W_1, ..., W_n)$

We use a Bayesian belief network to combine various weighted conditional probabilities and to compute the likelihood of the projection. A Bayesian belief network is a directed acyclic graph (DAG) whose nodes are the variables or propositions and whose arcs represent the existence of influence and are quantified by the conditional probabilities (Pearl, 1988). The weighted conditional probabilities for historical pattern, continuity and forecast for a state variable such as wind can be combined based on a belief network assuming the independence between the variables. Finally the likelihood measures for all state variables can be combined in a similar fashion to produce the final output of the likelihood measure for a given projection of oil slick movement.

4 IMPLEMENTATION

The intelligent system for predicting oil slick movement described in this paper is currently under development. Major issues in the implementation are discussed in this section.

First, there are various types of knowledges required in this system. A representation scheme that can modularize knowledge base by encapsulating knowledge of the same type in a relative independent unit is very desirable, since it would facilitate the system operation and maintenance. Secondly, a large amount of data is used to acquire different types of knowledge in this system. A method to provide uniform treatment of data and knowledge could facilitate the maintenance of the system, promotes the reuse of system functionalities and thus reduce the development cost.

An object-oriented paradigm is adopted to address the above issues and to implement the described system. We use an object-oriented representation for both data and high level knowledge. Related data and the knowledge are combined into a single unit, i.e., an object. Since data and other types of knowledge are treated uniformly, it simplifies the development process and reduces the development cost.

The user interface is to be developed using X window based utilities. The user is prompted to provide initial data value such as current wind velocity, wind direction, etc., via an application window. Then there are optional windows for the user to enter expert's domain knowledge such as the judgement on the reliability of historical weather data for this particular spill area, the geographical characteristics that affect the spill movement and others. If no domain expert's knowledge is entered, some default knowledge is assumed. The output of the system is in both statistical and graphical form. A table would list all inputs and the corresponding outputs. Various most likely scenarios for the oil slick movement are overlapped on the same plane and the intensity of color would represents the relative likelihood of a scenario.

A prototype system is being implemented in C++ on Sparc workstation. Some realistic testing data has been collected for validate the system once the implementation is completed.

5 SUMMARY

The goal of this research is to build an intelligent system to predict the oil slick movement to support the decision making in the management of a marine oil spill. The problem is characterized by the high degree of uncertainty and impreciseness associated with the observational data and the operating environment. The key to produce a prediction that is useful to the decision maker in managing an on-going oil spill is to acquire and to incorporate into a prediction as much knowledge as possible to reduce the amount of uncertainty the decision maker has to face. The major contributions of this system are the integration of a knowledge base containing diverse types of knowledge with a set of mathematical models simulating various oil slick motions and the adoption of an object-oriented paradigm to address various implementation and knowledge representation related issues.

REFERENCES

J.W. Anderson. Characteristics of dispersion and water soluble extracts of crude and refined oils and their toxicity to estuarine crustaceans and fish. *Marine Biology*, 1974.

J.A. Fay. Physical processes in the spread of oil on a water surface. In *Proceedings of the Joint Conference on the Prevention and Control of Oil Spills*, pages 463–67. American Petroleum Institute: New York, 1971.

Theresa Paluszkiewicz and Charles F. Marshall. Comparison of techniques for forcing an oil spill trajectory model. In *Proceedings of International Oil Spill Conference*, pages 547–553, February 1989.

J. Pearl. *Probabilistic Reasoning in Intelligent Systems: Networks of Plausible Inference*. Morgan Kaufmann, 1988.

William B. Samuels, Norden E. Huang, and David E. Amstutz. An oil spill trajectory analysis model with a variable wind deflection angle. *Ocean Engineering*, 9(4):347–360, 1982.

J. Tsao, J. Wolter, and H. Wang. Model-based understanding of uncertain observational data for oil spill tracking. *3rd Intl. Conf. on Industrial Fuzzy Control and Intelligent Systems*, 149-154, 1993.

J. Tsao, J. Wolter, and H. Wang. Model-based explanation of uncertain observations. *Proceedings of 1994 IEEE International Conference on Systems, Man and Cybernetics*, 2578-2583, 1994.

S. Venkatesh. The oil spill behaviour model of the Canadian Atmospheric Environment Service. part I: Theory and model evaluation. *ATMOSPHERE-OCEAN*, 26(1):93–108, 1988.

S. Venkatesh. Model simulations of the drift and spread of the Exxon Valdez oil spill. *ATMOSPHERE-OCEAN*, 28(1):90–105, 1990.

D. Weld and J. De Kleer. *Readings in Qualitative Reasoning about Physical Systems.* Morgan Kaufmann, San Mateo, CA, 1990.

G. N. Williams, R. Hann, and W.P. James. Predicting the fate of oil in the marine environment. In *Proceedings of the Joint Conference on the Prevention and Control of Oil Spills*, pages 567–572. American Petroleum Institute: New York, 1974.

L. Zadeh. Outline of a new approach to the analysis of complex systems and decision processes. *IEEE Trans. on Sys., Man and Cybernetics*, 3, 1972.

PART EIGHT

Environmental Data Visualization

23

Data Animator: Visualization of two- and three dimensional data as computer generated animations

Efraim Halfon
National Water Research Institute
Canada Centre for Inland Waters, Burlington, Ontario, Canada L7R 4A6,Tel. 905-336-4917; Fax. 905-336-4972, E-mail: EFRAIM.HALFON @ CCIW.CA

Abstract

Data Animator V1.0 is a powerful, easy-to-use scientific visualization package. Its main purpose is to generate animations from any data set collected over time. Geographical references such as a shore, bathymetry information, etc., may be added for additional clarity. Visualization of data as animations can greatly simplify the interpretation of field measurements. Data Animator is designed (but not restricted) to display data collected in aquatic environments, lakes, rivers, estuaries, oceans, etc., in a clear, concise way using colours to represent ranges of data values. Data sets can also be displayed as static images (keyframes). A graphic user interface (GUI) allows the user to choose a viewpoint, fonts, colour palette, data and keyframes. Point-and-click mouse operations allow the user to manipulate many features, with immediate on-screen feedback. This GUI allows the user to quickly design an attractive animation (or single frame) to exact specifications. Animations are generated by defining 'keyframes' of known data, each located at a specific point in time. The program can then interpolate over time, between keyframes, to create smoothly animated transitions (in-between frames). Two types of graphs can be rendered with Data Animator. 'Plane' type graphs are horizontal slices at a depth specified by the user. 'Transect' type graphs are vertical slices along a straight line defined by the user. Data Animator can make use of both shore outline information and three-dimensional bathymetry information. This allows for the generation of realistic-looking graphs which follow the shape of the aquatic environment. A typical animation, in FLC format, can be displayed at 30 frames per second. Animations can be displayed on a computer monitor or transferred to video tape.

Keywords

Scientific visualization, computer graphics, animation, data analysis, lake, aquatic system, 3D, 2D, limnology, oceanography

1 INTRODUCTION

Data collected in three dimensions are commonly available from limnological research. A ship collects data at several stations, and at many depths. When data are collected over time at many locations, a four dimensional picture surfaces. The fourth dimension is temporal. A computer generated animation can visualize the dynamics over time of three dimensional data.

While in the medical field (Herman and Liu, 1979; Sunguroff and Greenberg, 1978; Kaufmann, 1991) computer imaging is important for diagnostics and planning operations, in limnology computer imaging might be important to visualize the movement of water masses, the development of algal blooms, episodes of oxygen depletion, etc. Advanced methods of computer graphics can help scientists and water managers alike (Halfon, 1989).

The purpose of Data_Animator is to display data as computer generated animations. The lake bathymetry and topography are also an integral part of the package. Thus, the main function of the program is to generate (render) computer animations from data. This main advantage of this program is that it is user friendly and it can be learned in about an hour. Other more comprehensive visualization programs can only run on expensive workstations with a long learning time, while Data Animator runs on PCs and its use can be learned in about an hour. Best performance can be obtained on Pentium based machines. Its animations can be viewed on the computer or transferred on video.

Scientific visualization (Halfon, 1995) is a new and rapidly expanding field of research. Visualization permits a dynamic comparison of data collected over a season or over several years. Understanding of the data is facilitated by this tool. Computer animation, computer graphics, and video technology can also be used to present both to the general public and to specialized scientific audiences research result on aquatic environments. The use of animation can be expanded to include preparation of interactive information tools, such as CD-ROMs. Data Animator (Halfon et al., 1993, 1994; Halfon and Howell, 1995; Halfon and Tartari, 1995) has been used to interpret temperature and oxygen data collected in Hamilton Harbour, Lake Ontario and Lake Orta, an alpine Lake in Italy.

Limnological data are quite sparse in space. While a catscan (a three dimensional x-ray) might assemble information with a size of 512 x 512 x 512 pixels within hours, a limnologist would be hard pressed in collecting data at tens of stations, and at many depths, in less than a few days of field work, plus laboratory time. Lack of data creates problems of interpolation and extrapolation when computer animations are created. Also, to display three dimensional data collected in a lake, the lake bathymetry must be included to show the relation of the data to the bottom, and to the shoreline. In computer graphics, this relation takes the form of a mask. No data are shown on land. The development of this three dimensional mask is computationally challenging.

1.1 The grouping facility

One very important option is what we call 'grouping'. Grouping allows the user to separate regions which are physically close but logically separate, such as two bodies of water connected only by a narrow channel. When grouping is needed the number of groups can be defined as two, or greater. The grouping facility allows the user to define regions of a graph to be interpolated independently from each other. This is done by simply drawing polygons on a map of the region. Points which fall inside a polygon are interpolated using only the data that are also found within the polygon. The grouping facility has been developed to independently render different areas. The rationale is that some areas are not collected in reality and therefore the interpolation process should only use some stations and not others. The word grouping has been chosen since it means to divide the area to be rendered in different groups. The first group by default is zero (this number is for internal use of the program only).

2 DATA REQUIREMENTS

A standard input format was designed to incorporate the data and its geographical and temporal reference. The format is ASCII. Data collected in Hamilton Harbour, Lake Ontario, are used here as an example. The first column is concentration of dissolved oxygen, the second column represents temperature data, the third column is depth from the surface, the fourth is conductivity, followed by transmissivity and pH. The order of the data in the column is immaterial. Depth is just another variable, thus is it possible to visualize, for example, oxygen vs. temperature, pH vs conductivity, etc. Data files must be organized by sample time. All samples taken within the same period are included in one file. The date and time are read from the first sample in the file. Each file can contain data from several sample sites.

Site Header
Site Data
"#EOF" String

The following is an example of a sample data file (90051):

```
 lat    long    date   station number

431828 794645 90 05 28 908
  13.4702  12.8028    .1666   55.3785
  13.4497  12.8020    .2265   55.6547
  13.8910  12.7128    .4615   59.3252
  13.6445  12.5669    .6537   59.8703
  13.4717  12.5122    .8816   58.6658
  13.4601  12.3814   1.0810   56.7397
  13.4271  12.3054   1.2946   55.8208
     .       .        .       .
     .       .        .       .
     .       .        .       .
     .       .        .       .
#EOF
```

A shoreline file may contain as many shoreline objects as needed. Objects may include islands, rivers, several lakes, etc.

3 GRAPHICAL USER INTERFACE (GUI)

The Data_Animator user interface (GUI) was designed as drop down menus, each menu designed for data input, rendering of key frames, keyframe animations or full animations. The GUI also lets the user choose the topography and the bathymetry files to use, the fonts to use for titling, the number of colours (up to 256), the colours and the range of data each colour represents. Point-and-click mouse operations allow the user to manipulate many features, with immediate on-screen feedback. This GUI allows the user to quickly design an attractive animation (or single frame) to exact specifications.

KeyFrame-Oriented Construction of Animations

Animations are generated by defining 'keyframes' of known data, each located at a specific point in time. The program can then interpolate over time, between keyframes, to create smoothly animated transitions (in-between frames).

Animations are created in much the same way as movies are. Snapshots of action, called 'frames', are recorded at regular intervals in time. They are then displayed quickly one after the other to create the illusion of continuous movement. Computer animations are quite similar: they are a collection of still images of the computer screen which can be displayed very quickly, one after the other.

To create the illusion of smooth movement, the frames of an animation, or movie, must be displayed very quickly. For example, standard North American television screens display 30 still frames every second. This is also the standard rate for displaying animations on the Amiga computer. For example, an animation composed of 1800 still frames will play for one minute at the standard rate of 30 frames per second (fps).

This high rate of display presents a problem for visualizing scientific data. In most cases, the data provided is not nearly dense enough in time to generate such a large number of frames. Because of this, a method has been developed to 'stretch out' the data provided to the program.

Frames are categorized into two different types: KeyFrames, and in-between frames. 'KeyFrames' are those frames where known data values are provided by the user. These keyframes are created and manipulated by the user in the Data Animator program. A simple animation of only keyframes can be generated by the program, and if played back very slowly (eg. 1 fps) it will provide a short, jumpy display of the data (like a slide show).

The 'in-between' frames are generated by the computer by interpolating over time between consecutive keyframes. Generation of in-between frames is the time consuming part of creating an animation. Each pixel on the screen has to be generated by linear interpolation in time (interpolation in space is used for key frames). In countries with the NTSC standard (e.g., North America and Japan), televisions display 30 frames per second, thus, to create an animation where each frame was rendered separately, an 1800 frames animation had to be developed (Of course it is also possible to play an animation at a slower rate, so that each NTSC frame corresponds to more than one rendered frame). To compute the number of in-between frames a simple proportion was made to accommodate these boundaries. In this way, the computer can insert frames of computed information between the keyframes of known data to create longer, smoother animations. The following options are available in Dat Animator:

Option	Description

Single Frame	To render only the current keyframe.
Key Frame	To render multiple keyframes to an FLC file. When the 'Begin Rendering' command is issued, rendering will begin at the first keyframe and continue for the number of frames specified in the 'Number of Frames' input box.
Full	To render a full animation, including keyframes and interpolated in-between frames, to an FLC file. The total number of frames rendered will match the value in the 'Number of Frames' input box. The number of in-between frames rendered between each pair of keyframes will be calculated based on the sample dates of the keyframes.

The GUI contains many commands to control the graphical settings.

1. Graph type
   - plane, transect, 3D

2. Graph limits
   - north, south east and west limits in degrees.minutes.seconds of latitude/longitude
   - surface and bottom in meters

3. Graph view
   - scale factors for each axis

4. Display options
   - bounding box
   - site markers, site numbers
   - label depth axis
   - shoreline

5. Interpolation
   - Closest-Sites interpolation, as well as the number of closest sites to use in the interpolation calculations

6. Graph labels
   - main title and colour legend title
   - Date display options (any combination of day, month, year)
   - Fonts for all labels. For video purposes, we recommend choosing a large and easily legible font.

7. Colours
   - number of colours to use (256 available)
   - lowest data value of lowest colour
   - data range for each colour

8. Animation type
   - single frame, key frame or full animation
   - number of frames to render

9. Transect line
   - straight line defined by the user along which a transect is rendered. A transect line can be interactively chosen by moving the mouse on the screen to a location, start of the transect, and then to another to indicate the end of the transect. The computer program then identified the closest stations to each point in the transect line to create the desiderate visualization. The number of closest stations can be chosen by the user, usually this number ranges between two and five. This user-definable feature gives the program flexibility to deal with different data set densities. The closest stations, weighted by the inverse of the distance squared, are also used in the top view. Thus, all data visualizations are internally consistent.

Each menu command has an associated requester to change settings. If possible, any changes are reflected immediately in the main project window.

4 ANIMATIONS

Data Animator uses two interpolation methods and one assumption are used to create the animation. Vertical interpolation is done using a spline. Horizontally, the interpolation is linear, but non-linearly modified by the weights (inverse to the square of the distance from the station, as mentioned in the previous paragraph). Once the variable of interest is computed, a colour is assigned to each pixel to represent the variable at that location. A colour map on the right of the screen, describes the relation between the values of the variable and the colour. The user can decide the colour palette, i.e. which colour represents which range of data.

Figure 1 shows four frames out of an animation showing the behaviour of chlorophyll in Cootes Paradise from August 6 to August 25, 1986. To represent these frames in black and white rather than colour or grey scale, these frames have been edited for clarity (The original colour pictures and the full animation for 1986 are included in the ISESS1.ZIP file; The animations, in FLC format, can be obtained through anonymous FTP: ftp.cciw.ca, directory incoming/halfon). Animations for all water quality parameters (Total phosphorus, suspended solids, Secchi disk, etc.) for all subsequent years have also been performed. The interesting observation from this sequence is the presence of a gradient from west to east and a gradient from east to west, stations 1 to station 3. Indeed station 3 has a concentration of chlorophyll lower than at stations at the west and the east. The animation show a dilution pattern originating in Hamilton Harbour and moving to the west. Conversely the damaging effects of the wastewater treatment plant in the west end stop in the west pond and do not influence the main part of

Cootes Paradise. This sequence of events in 1986 is not immediately evident from the spreadsheet data, only a dynamic visualization of the data permitted this interpretation.

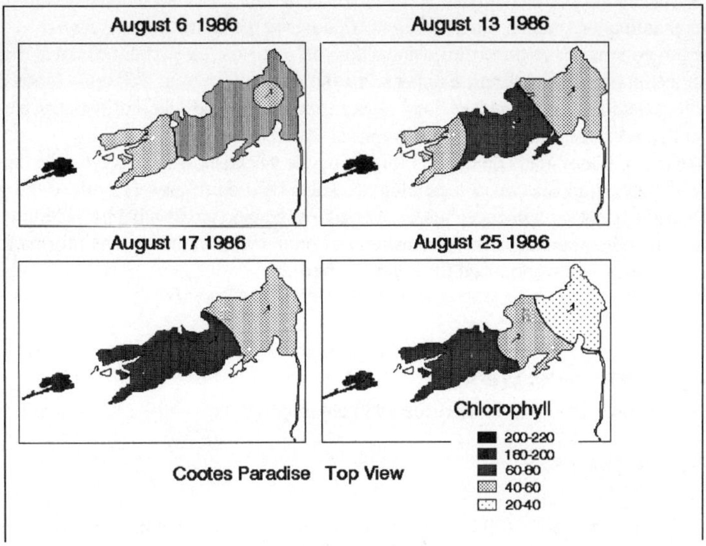

Figure 1 Four frames of the animation of chlorophyll concentration in Cootes Paradise, Lake Ontario are presented. The forth frame is a keyframe, i.e., data were used to create this frame. The other three frames are in-between frames created with interpolations in space and in time. The full colour animation is in FLC format and is posted (see text).

5 DISCUSSION

In Cootes Paradise data have been collected for about a decade, at several stations a few times a year. The amount of data is large considering the fact that at each station many water quality variables were collected. Visualization permits a dynamic comparison of data collected over a season or over several years. Several animations can be run in different windows at the same time. Two dimensional displays are projections of the three dimensional data on one axis. A two dimensional display might be a transect or a top view. Visualization implies that the data collected at different stations and different times are interpolated in space and in time. Lake topography and lake bathymetry files are used to relate data collected in the lake(s) with topographical features.

Data_Animator was initially developed for use by the scientists who collected the data. Computer animation, computer graphics, video technology can also be used to present both to the general public and to specialized scientific audiences research result on aquatic environments. The use of animation can expand to include other scientific information and preparation of interactive information tools. The program runs on PCs and is optimized for Pentium chips.

6 ACKNOWLEDGMENTS

Dr. Pat Chow-Fraser kindly provided the 1986 chlorophyll data shown in Figure 1.

7 REFERENCES

Herman, G.T. and Liu, H.K. (1979) Three dimensional display of human organs from computed tomograms. *Computer Graphics and Image Processing,* **8**, 1-21.

Halfon, E. (Ed.) (1989) Microcomputers in ecological modelling: a special issue dedicated to research, education and computer graphics, *Ecological Modelling*, Vol. 47.

Halfon, E., Tartari, G., Brendon, D., and Binelli, A. (1993) Visualization and limnological interpretation of liming treatments in a deep lake, Lake Orta, Italy, using computer generated animations. In IFIP, WG 5.11 Working conference on Computer Support for Environmental Assessment, CSEIA '93, villa Olmo, Como, Oct. 6-8, 1993. pp. 82-91.

Halfon, E., Tartari, G., Howell, M., and Binelli, A. (1994) Visualization of pH changes due to liming treatments in Lake Orta, Italy, using computer generated animations. In G. Guariso and B. Page (eds) Computer Support for Environmental Assessment, Elsevier Science, 37-48.

Halfon, E. and Howell, M. (1995) Visualization of limnological data as two- and three-dimensional computer generated animations. *J. Biol. Systems* , **4**, 3-25.

Halfon, E. and Tartari, G. (1995) Visualizzazione di dati raccolti nel Lago d'Orta negli anni 1989-1991. *Acqua e Aria* (in press).

Halfon, E. (ed) (1995) Special issue on the use of computer graphics and visualization. *J. Biological System* s, Vol. 2.

Kaufmann, A. (1991) *Volume Visualization* , IEEE Computer Society Press, Los Alamitos, Ca., 487 pp.

Sunguroff, A. and Greenberg, D. (1978) Computer generated images for medical applications. *Computer Graphics*, **12**, 196-202.

8 BIOGRAPHY

I am a research scientist with Environment Canada. I am also a professor in the Biology Department at McMaster University in Hamilton, Ontario. My interests are in scientific visualization, mathematical modelling of the fate of toxic contaminants, risk assessment of toxic contaminants and their ranking.

24

Visualizing the Spatial and Temporal Dynamics of Ozone Concentration Data

H.F. Mayer and W. Haas
JOANNEUM RESEARCH, Institute of Information Systems
Steyrergasse 17, A-8010 Graz, Austria
e-mail: {mayer,haas}@pbox.joanneum.ac.at

J. Züger and W. Loibl
Austrian Research Centre Seibersdorf
A-2444 Seibersdorf, Austria
e-mail: {zueger,loibl}@zdvaxa.arcs.ac.at

Abstract

In Austria the ozone concentrations are measured by a network of more than 130 monitoring stations. In order to perceive the temporal and spatial dynamics of such concentration data, it is necessary to develop special visualization techniques. This paper describes the various steps in preparing the data and providing results on graphic workstations, on video tapes and online through the World-Wide Web.

Keywords

Environmental research and planning, ozone, visualization, World-Wide Web

1. INTRODUCTION

Research into the origin, distribution and effects of ozone is internationally regarded as one of the biggest challenges in the area of environmental research and planning. This particularly applies to the high levels of ozone in the tropospheric boundary layer in summer which are a major environmental concern in Austria. A network of more than 130 monitoring stations has been established throughout the country to measure and analyze ozone concentrations. This network is operated by the Austrian provinces and the Federal Environment Agency. In the past few years the Austrian Research Centre Seibersdorf (Loibl, 1993) developed a model to estimate the spatial distribution and diurnal variation of ozone referring to the point data measured every half an hour at the monitoring stations. By using a digital elevation model and the monitoring data, the ozone concentration of areas between monitoring sites can be calculated.

The results of this spatial interpolation procedure are ozone concentration maps for every half an hour of a day. A fundamental disadvantage of such static visualizations is that the temporal sequence and dynamics of the processes behind it can only be grasped incompletely. To clearly depict the dynamics of ozone development for both the analyzing reseacher and the interested public, we applied interactive and dynamic visualization techniques. The results of this piece of research are documented on video and will be discussed in this paper.

2. OZONE CONCENTRATION AS A FUNCTION OF TIME AND ALTITUDE

Monitoring data are only representative for the pollution levels at the very monitoring sites. However, information about the spatial distribution of ozone levels is necessary to assess the impact of pollution on human health and ecosystems. Therefore, the ozone levels in the areas between the monitoring stations have to be calculated.

Ozone is produced from precursor air pollutants under sunlight. The rate of formation is dependent on meteorological and chemical conditions. In areas with anthropogenic pollution there is a build-up of ozone with increasing solar radiation during the day, and a reduction of ozone at night. In higher altitudes, ozone levels are higher first and remain much the same throughout the day.

The diurnal variation of ozone is bell-shaped in valleys and more uniform with higher altitudes. So in regions with complex terrain and pronounced topographical variation - as it is the case in Austria - ozone levels will be very different even within small areas.

The Austrian Research Centre Seibersdorf (Züger, 1994) has developed a method for the spatial interpolation of ozone monitoring data. This method is based on a mathematical function which describes the altitude and diurnal variability of ozone. Using a digital elevation model and the monitoring data, the ozone concentrations can be calculated for areas between monitoring sites. After the preprocessing of this data the concentrations are available as half-hour mean values on a rectangular grid with a resolution of 2500 x 2500 meters.

3. STATIC VISUALIZATION

Before the temporal animation, which is produced by single frames following one another, the preprocessed data have to be mapped onto suitable geometric objects (Brodlie, 1992). These objects can be viewed on various output media (monitor, paper, video), using different rendering methods.

In our case the relationship between ozone levels and relative altitude was to be documented. For this purpose we generate a three-dimensional grid of a digital elevation model of Austria in the resolution of 2500 x 2500 meters. Each grid point is colored according to the concentration at this position. The color in the areas between the grid points is interpolated by the rendering process using Gouraud shading (Foley et al, 1990).

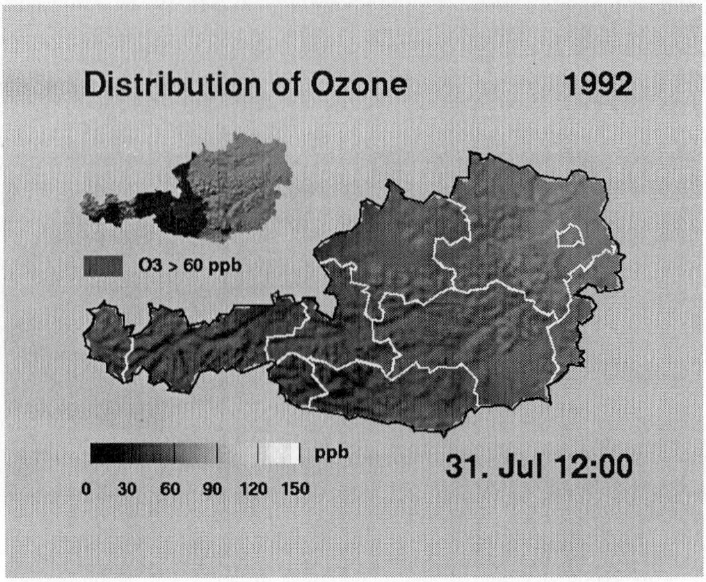

Figure 1 Ozone concentrations on 31 July 1992, 12:00 a.m.

The selection of a suitable colormap for the coding of the data is of crucial significance. There is some literature on this topic (Hopgood, 1991 and Murch, 1986), however, the results are individually different. For that reason we carried out a survey with approximately 50 persons who were asked to sort some color cards according to their subjective perception of the severity of pollution. The results of this survey were analyzed with statistical methods (cluster analysis) and delivered two representative color maps which were incorporated into the map used for the actual visualization.

The elevation model is overlayed with polylines representing the federal and provincal borders to help the viewer get his bearings.

The mapping of the data onto three-dimensional geometries allows studying from different viewpoints, local zooming and flying over regions in an interactive visualization environment (e.g. graphic workstations). The elevation model is illuminated from north-west to make the height influence visible from all perspectives.

4. LAYOUT OF CONCENTRATION MAPS

After selection of the principal visualization method, the overall graphical representation has to be constructed. The goal is to visualize all information in a highly expressive way. However, caution is needed for the combined visualization of several scalar or vectorial data in order not to confuse the observer. This principle particularly holds for animations where the observation time is usually fixed.

Considering all these findings, we selected the following layout for the visualization (see Figure 1):

- The ozone concentrations of Austria are shown in detail on a nearly screen-sized map viewed from above.
- Two states of pollution colored green and red are indicated on a small map of Austria, thus specifying which areas have got a concentration level below or above a half-hour mean value of 60 parts per billion. This value is an Austrian guideline for human health protection.
- On the bottom right there is the corresponding date and time of the visualized data set.
- On the bottom left the color map for the coding of the big map is displayed.

Other variants of visualization, e.g. perspective view of the map from the side, proved to be interesting. For the final version, however, we chose the simple view from above to ensure a high degree of information and to facilitate analysis with a kind of visualization the observers are used to.

5. DYNAMIC VISUALIZATION

After choosing the layout and visualization methods for a single map, the dynamic visualization can be built by concatenating a sequence of single frames of successive data sets. For that reason the concentration data have to be interpolated for every moment between two given half-hour data sets. We use a simple linear interpolation for this task. Higher order interpolation methods proved to be not necessary for this kind of problem.

1250 frames are needed to play back one day of an ozone episode speeded up in 50 seconds. As high-performance graphics workstations are not yet capable of producing such an animation in real-time, each frame has to be calculated by its own and recorded to a digital video disk recorder. As soon as all frames are recorded the animation can be viewed in real-time and copied to the usual magnetic tapes.

Also the small bandwidth of video has to be considered when producing animation on video. In particular there are problems with fine details (e.g. thin lines) and with a limited color spectrum.

A script was developed to integrate the visualization together with a short description of the spatial interpolation method into a self-describing and entertaining video. This video was presented to a broad audience within the framework of a fair for environmental technology.

6. ONLINE VISUALIZATION IN THE WORLD-WIDE WEB

Apart from producing a video where the information is fixed to a chosen ozone episode, we developed a concept to provide access to online information. The emerging World-Wide Web (Berners-Lee, 1992) offers an open environment for implementing such multi-media applications and to distribute the information to everyone on the Internet.

Animations are provided in the form of MPEG (ISO 11172, 1993) video streams, an international standard for compressed digital video. These streams can be decoded by software and viewed on PCs and workstations. For playback in realtime, however, a hardware MPEG decoder is necessary. Apart from animation it is possible to point and click at the monitoring stations on a map to obtain the latest concentration data.

A prototype of this application can be reached by accessing the address (URL) `http://iis.joanneum.ac.at/ozonanim`.

7. CONCLUSION

The spatial and temporal dynamics of ozone concentration data can only be visualized by using interactive and dynamic visualization techniques. The results of such visualizations can be analyzed on graphic workstations, presented to a broad audience on video or made available online using the World-Wide Web. In all cases, these techniques have proven to be very helpful, either to introduce to the problem in short time or to help grasp the problem and possible solutions.

ACKNOWLEDGEMENTS

The work described in this paper was done in cooperation of JOANNEUM RESEARCH and the Austrian Research Centre Seibersdorf. It was supported by the Austrian Federal Ministry of Science and Research. We would like to thank the Federal Environment Agency of Austria who provided the ozone data.

REFERENCES

Berners-Lee T.J, Cailliau R., Groff J.-F. and Pollermann B., "*World-Wide Web: An Information Infrastructure of High-Energy Physics*", Proceedings of Software Engineering, Artificial Intelligence and Expert Systems of High Energy and Nuclear Physics, La Londe-les-Maures, France, 1992.

Brodlie et al. (Eds.), *Scientific Visualization*, Springer, 1992.

Foley J.D., van Dam A., Feiner S.K. and Hughes J.F., *Computer Graphics - Principles and Practice*, Addison-Wesley, 1990.

Hopgood F.R.A., "*Using Color in Computer Graphics*", AGOCG Technical Report 4, 1991.

ISO/IEC International Standard 11172, *Information Technology - Coding of moving pictures and associated audio for digital storage media up to about 1.5 Mbit/s*, 1993.

Loibl W., Züger J., Kopsca A., "*Flächenhafte Ozonverteilung in Österreich für ausgewählte Ozonepisoden 1991*", in German, Report of the Federal Environment Agency Austria, UBA-93-071, 1993.

Murch G.M., "*Human Factors of Colour Displays*", Advances in Computer Graphics, Springer, 1986.

Züger J., "*Erstellung von Datensätzen zur Animation des Ozonkonzentrationsverlaufs von Ozonepisoden 1992*", in German, Report of the Austrian Research Centre Seibersdorf, OEFZS-A-3255, 1994.

BIOGRAPHY

Harald F. Mayer graduated in technical mathematics from Graz University of Technology in 1989 with a degree dissertation in the field of digital image processing. Since 1990 he has been working for the company of JOANNEUM RESEARCH where he focuses on research projects dealing with scientific visualization and computer graphics. His activity centers on the visualization of results from Finite Element simulations (pollution transport), Computational Fluid Dynamics (turbulent flow in engines) and ozone concentration data.

Werner Haas graduated in technical physics from Graz University of Technology in 1973. Afterwards he worked at the Institute for Hydraulics at the Graz University of Technology on a computer-aided measurement system. Subsequently, he was employed by a software house where he carried out extensive developments on Finite Elements models and methods applied to hydraulics and mechanics. Since 1988 he has been working for the company of JOANNEUM RESEARCH where he is now the director of the Institute of Information Systems.

Johann Züger has been working on air- and groundwater pollution dispersion models since several years. During last year he was dealing with photochemical models to simulate ozone built up deputy on precursors and additionally he is working on animated spatial results of the maintained models.

Wolfgang Loibl was the leader of several projects dealing with the modeling of spatial patterns of diurnal ozone variation in alpine regions depending on day-time and altitude. Further projects were spatially disaggregated emission inventories (NMVOC, NO_x, Methane, etc.) using land cover patterns to model the emissions density with spatial resolution and estimations of critical loads and levels of this and other pollutants depending on land cover.

25

VOLUME VISUALIZATION OF WATER QUALITY DATA IN HAMILTON HARBOUR, LAKE ONTARIO

Efraim Halfon
National Water Research Institute
Canada Centre for Inland Waters, Burlington, Ontario, Canada L7R 4A6,
Tel. 905-336-4917; fax 905-336-4972; E-mail: efraim.halfon@cciw.ca

Abstract

Visualization in the physical and natural sciences is distinguished by the need to deal with data sets that are volumetric, time varying and multi-variant. The goal of this project is implementation of a system whereby researchers can access, analyze, and visualize water quality data collected in three dimensions (geographical co-ordinates and depth). Using computer graphics it is now possible to combine data, their three dimensional location and lake topography to create images of water quality patterns which supersede conventional surface, two dimensional, graphics (GIS). Through solid modelling, oxygen and temperature data collected on May 28, 1990 and August 8, 1990 in Hamilton Harbour, Lake Ontario, are mapped into voxels and visualized on a two dimensional screen. The calculation of the 3D representation allows the accurate computation of volumetric properties since each voxel has water quality values associated with it and these values can be summed or elaborated numerically as needs arise. For example the lake has a volume of 253,947,500 m^3, the water mass (May 28, 1990) at 12-13 °C had a complex three dimensional shape with a volume of 61,036,250 m^3 and on August 8, 66% of the lake volume had oxygen levels over 4 mg L^{-1} and suitable for fish growth, contrary to expectations. A third benefit of visualization is that the data can be viewed interactively from different viewpoints thus increasing the interaction between scientist and the data.

Keywords

Volume, visualization, lake, aquatic, 3d, Hamilton Harbour, Lake Ontario, temperature, oxygen, water quality, bathymetry, graphics

1 INTRODUCTION

This project investigates and compares different approaches to visualization of sampled volumetric data collected in lakes and in particular in Hamilton Harbour, Lake Ontario. The interpolation of 3D data, the mapping from numeric data to visual parameters such as colour and transparency are discussed.

Encouragement to more fully utilize time-varying, multi-dimensional data from Hamilton Harbour and other areas of concern is a further objective.

Hamilton Harbour is located at the western end of Lake Ontario, it has a triangular shape with maximum dimensions eight kilometres from east to west and five kilometres from north to south; the maximum depth in the middle of the bay is about 26 metres and an additional deep area is located at the east end. A ten metre deep channel connects it to Lake Ontario. Hamilton Harbour has been widely studied in the past ten years (Charlton, pers. comm.) since it is one of the 42 areas of concern within the Great Lakes and furthermore the International Council for Local Environmental Initiatives has designated the city of Hamilton as Canada's model city under the United Nations Agenda 21 program (ICLEI, 1990).

The Hamilton Harbour Research project will have archived over a gigabyte of heterogeneous image data by the end of this year. Other sites are similarly amassing large, complicated data sets which ecologists want to transform into useful information to provide new insight into long-term ecological phenomena. Unfortunately, the procedures, software tools, and user friendly interfaces necessary to facilitate access to these data sets are currently lacking which precludes analysis and subsequent visualization by ecologists.

In traditional (pre 1970) computer graphics, polygons and lines (e.g., a wireframe) represented three dimensional (3D) volumetric objects but some 3D data sets, e.g., the volumetric water quality patterns collected in a lake, might not consist of surfaces and edges at all. As a result, Greenleaf *et al.* (1970) were the first to introduce new methods of visualization to extract information from volume data. This process called *volume visualization* is defined by Kaufman (1992) as "a **direct** technique for visualizing volume primitives without any intermediate conversion of the volumetric data set to surface representation."

The first applications were in medical imaging and they are still dominant (e.g., Adams *et al.*, 1990; Rhodes *et al.*, 1987). McCormick *et al.* (1987) stated that volume visualization was emerging as a new visualization environment in scientific computing; for example Hibbard and Santek (1989) presented interesting uses in the earth sciences.

Kaufman (1991a) provided an insight into this technology stating that "the objective of volume visualization is to peer inside the volumetric objects to view that which is not ordinarily viewable and to probe into the voluminous and complex structures and their dynamics to comprehend that which is not ordinarily comprehensible." Halfon *et al.* (1993), Assel *et al.* (1994), and Halfon and Howell (1995) recently presented two dimensional computer visualizations of limnological data. Here, a framework is described to visualize and interpret 3D data. In this framework there are three steps: data collection, solid modelling or transformation of the data in a specific volumetric format, and finally volume rendering or the visualization process. For general applications consult Kaufman (1991b), Foley *et al.* (1991) and Bergeron and Kaufman (1994).

2 THE VISUALIZATION PROCESS

Data

Water quality data (water temperature, oxygen concentration, conductivity, pH, etc.) have been collected by Charlton (pers. comm.) every year since 1985 at 26 stations (the number of stations has slightly increased or decreased over the years).

Solid modelling

This area of computer graphics has arisen following the need for modelling objects as solids (Foley *et al.*, 1991). A lake is a volumetric object and each point in the lake has the value of a water quality variable (e.g., temperature, pH, etc.) associated with it. The "point", or cell, in the lake is of course related to the resolution used in the computer representation and the cells are a collection of adjoining nonintersecting solids. When the cells are equal and arranged in a fixed, regular grid, they are called **voxels** (volume elements), analogous to pixels. Each voxel is a quantum unit of volume and has a numeric value (or values) associated with it that represents some measurable properties) and the 3D volumetric data set resides in an **integer** grid of voxels called a *cubic frame buffer.*

The visualization process consists of several steps necessary to organize the data in the proper voxel format: A first step is to divide the area of interest into a grid; the grid chosen for Hamilton Harbour is 50 x 50 metres. The depth value in each grid unit was obtained by interpolation using the SAS contouring program and the bathymetric data, which were originally in sparse form, were organized into a fixed grid.

Following the grid pattern, each voxel was defined as a rectangular parallelepiped (which allows for resolution in the different axes) with dimensions 50 metres x 50 metres by 0.5 metres and the cube frame buffer has dimensions of 200 x 127 x 52 for a total of 1,320,800 voxels. Over 2,500 bathymetry data and 14,000 shoreline data are available from the Canadian Hydrographic Service. The choice of the number of voxels and their size is left to the user for several reasons: As the number of cells increases, grid size diminishes, the ambiguity between land and water voxels decreases, but it does not disappear, unless the lake is a perfect cube (unlikely). However, the larger the number of voxels the larger is the computer memory requirement and the computing time.

The second step was to identify which voxels represent land (depth value of zero) and which water (grid depth value greater than zero). Out of the 1,320,880 voxels only 204,034 are water given the conical shape of the harbour. A 3D array was created where each land voxel was assigned a value of zero and each water voxel a value of one. A similar procedure was followed with the water quality data. Since voxels have dimensions of cubic metres and data collected along the depth axis have already a metric dimension associated with them (data collected at .1, .2, ..., etc. metres), station locations were converted from geographical to UTM coordinates [metres] before the interpolation was performed.

Interpolation of the water quality data from scattered stations to a uniform grid was performed using water quality parameter data collected at the nearest three stations. Weight factors were inversely proportional to the square of the distance from the three closest stations. Interpolation was performed in two dimensions, one layer at the time. Interpolation in the vertical axis was performed using the IMSL (1987) subroutine SCAKM which uses a spline function to interpolate the data. A spline interpolation could not be used in the horizontal plane since the IMSL subroutine SURFER could not be constrained to have the interpolated data stay within the observed range. The interpolation problem, that is fitting the data collected at arbitrarily located positions to a uniform grid is complex. While this problem has not been fully solved, Foley *et al.* (1993) have reviewed past work and proposed solutions.

The third step was to join the voxels with their water quality values through Boolean algebra, this process has been defined by Manley and Tallet (1990) as "clipping."

Volume rendering

Kaufman (1991) states that "to visualize the volume data set, the voxels can be projected into 2D pixel space and stored as a raster image frame in a frame buffer. This process, which is termed volume rendering (Drebin *et al.*, 1988; Frenkel, 1989; Levoy, 1988; Upson and Keeler, 1988), involves both the viewing and the shading of the volume image." The definition of Foley *et al.* (1991) is that "volume rendering is the process of displaying scalar fields," where a scalar field is a collection of all the numbers associated with each point in a volume. In summary "volume rendering is a direct display of volume primitives without any intermediate conversion of volume data to surface representation." (Kaufman, 1991a). There is no a priori assumption that "the data consist of tangible surfaces that can be extracted and visualized." Specialized algorithms have been developed to render clouds, humans and living objects (Blinn, 1982; Herman and Udupa, 1983). The application of volume rendering techniques to limnology is therefore not complex, but it requires the data preparation explained above.

Volume viewing and shading
Kaufman's (1991b) book can be consulted for a comprehensive review to create 2D projections from the 3D volumetric data, as well as volume rendering and shading. For this application the Hamilton Harbour voxel values were stored in ASCII format and the data set was read into memory by the Spyglass® Slicer[1] program (Spyglass, 1994). Slicer is a volume visualization software that can be used for scientific visualization; while the program focuses on the 3D primitives, it can also visualize the data using more traditional 2D primitives, such as surface rendering as slices and isosurfaces (three-dimensional equivalent of contour lines).

Surface rendering is defined as "an *indirect* technique used for visualizing volume primitives by first converting them into intermediate surface representation and then employing conventional computer graphics techniques to render them to the screen." (Kaufman, 1992). An isosurface (Levoy, 1988) is a surface where the data with the same values are interpolated together, for example a thermocline can be considered a three dimensional temperature isosurface. The surface can be rendered as a transparent or solid surface. The isosurface can also be coloured according to the values of the scalar field at the surface points (Mercurio, 1991).

3 RESULTS

Water temperature and oxygen concentrations
Water temperature on May 28, 1990 ranged from 11.2 to 18.9 °C and oxygen concentrations ranged between 3.8 and 15.5 mg L^{-1}; water temperature on August 8 ranged from 12 to 24 °C and oxygen concentrations ranged between 0 and 10.7 mg L^{-1}. Figures related to this paper can be found in the anonymous FTP site (ftp.cciw.ca in the directory incoming/halfon). The file name is isess2.zip. This file contains GIF files that show the visualization of the temperature data collected on May 28, 1990 under different rendering options. A more comprehensive set of figures can be found posted on the website http://www.cciw.ca/glimr/data/limnology/intr.html.

In May the bay was heating from the west while temperature at depth was about 11 °C. One interesting feature is a convective chimney extending from the water surface to a depth of about ten metres: this water mass has a temperature of 13 to 14 °C.

The circulation of the bay is complex and Hamblin (pers. comm.) is now developing a two dimensional mathematical model of this flow field: At the surface, currents are mostly from west to east following the prevailing winds with return flow at deeper levels. At the east end of the lake surface waters cool by mixing with Lake Ontario water that enters the bay from a channel and form a

[1] Spyglass is a registered trademark of Spyglass, Inc.

downwelling. Thus, just under the surface along the main east-west axis there is a current from west to east[2].

In August 1990 the 13 to 14 °C and 16 to 17 °C water masses are found at lower depths than in May, at about 15 and 10 metres, respectively. The water mass at the east end of the bay, > 23 °C, has the same surface temperature as the western part of lake Ontario.

Volumetric properties

The 3D representation allows computation of several volumetric properties: even if the bathymetry is uneven, the total lake volume can be easily computed by multiplying 231,053 voxels by the volume of each voxel (50 m x 50 m x 0.5 m or 1250 m^3) for a total of 253,947,500 m^3. Other properties, laborious to calculate through standard two dimensional projections (since these water masses have complex three dimensional shapes) are, for example, the water volume at a given temperature (Table 1) and its heat content. On May 28, the median volumetric temperature is 13 °C while on August 8, the median temperature is about 20 °C. The volumetric frequency distribution of temperatures is also very different on these two dates (Table 1). On August 8, 1990 the 13 to 14 °C and 16 to 17 °C water masses are found at lower depths than on May 28, at about 15 and 10 metres, respectively. The water mass at the east end of the bay, > 23 °C, has the same surface temperature as the western part of Lake Ontario.

Table 1 Heat content of Hamilton Harbour on May 28, and August 8, 1990. Each voxel has a volume of 1250 m^3

May 28, 1990

Temperature Range °C	No. of Voxels	Volume of Voxels [m^3]	% of lake volume [J]	Heat Content
11-12	51228	6.40 10^7	25.22	2.08 10^{15}
12-13	48829	6.10 10^7	24.03	3.19 10^{15}
13-14	39688	4.96 10^7	19.54	2.80 10^{15}
14-15	29608	3.70 10^7	14.57	2.24 10^{15}
15-16	14590	1.82 10^7	7.18	1.18 10^{15}

[2] Conductivity values in May 28, 1990 ranged between 550 and 760 with most of the water in the range 655-680 while Lake Ontario waters have an average value of about 300 (Charlton, pers. comm.).

16-17	12942	1.62 10^7	6.37	1.12 10^{15}
17-18	5775	7.22 10^6	2.84	5.28 10^{14}
18-19	497	6.21 10^5	0.24	4.80 10^{13}
				1.42 10^{16}

August 8, 1990

Temperature Range	No. of Voxels	Volume of Voxels [m^3]	% of lake volume [J]	Heat Content
12-13	10043	1.26 10^7	4.94	6.56 10^{14}
13-14	22203	2.78 10^7	10.93	1.57 10^{15}
14-15	15410	1.93 10^7	7.58	1.17 10^{15}
15-16	11474	1.43 10^7	5.65	9.30 10^{14}
16-17	11711	1.46 10^7	5.76	1.01 10^{15}
17-18	9279	1.16 10^7	4.57	8.48 10^{14}
18-19	6690	8.36 10^6	3.29	6.47 10^{14}
19-20	8083	1.01 10^7	3.98	8.23 10^{14}
20-21	8511	1.06 10^7	4.19	9.11 10^{14}
21-22	13119	1.64 10^7	6.46	1.47 10^{15}
22-23	50167	6.27 10^7	24.69	5.89 10^{15}
23-24	31019	3.88 10^7	15.27	3.80 10^{15}
24-25	5451	6.81 10^6	2.68	6.97 10^{14}

A similar calculation can also be performed for oxygen. About 34% of the lake had oxygen levels below 4 mg L^{-1} on August 8, and conversely, 66% of the harbour had waters suitable for fish growth. On May 28, oxygen depletion does not start at the bottom, as it could be expected, but it starts at the surface at the southern shallow part of the lake and it evolves northward in the form of chimneys.

4 DISCUSSION

Visualization in the physical and natural sciences is distinguished by the need to deal with data sets that are volumetric, randomly spaced, time varying and multi-variant. With this methodology limnologists can access, analyze, and visualize water quality data collected in three dimensions (geographical co-ordinates and depth). Manley and Tallet (1990) noted that "the ability to visually manipulate and gain quantitative information from three dimensional models[3] provides more information to the researcher over a significantly shorter period of time that a myriad of cross-sections and planar maps." The process of collating the data also produces information since the properties associated with each voxel are suitable for volumetric analysis. With the development of fast personal computers and properly designed software, it is possible to visualize limnological data on a personal computer quickly once the data have been interpolated and saved in a format suitable for volume visualization. This time frame contrasts with the months needed to collect a proper data set.

---

[3] Manley and Tallet (1990) used both mathematical models as well as data to understand the system they were studying.

Charlton (pers. comm.) has collected water quality data in Hamilton Harbour since 1985: After nine years of sampling, the amount of data available for analysis is quite extensive, since data are collected seven or eight times a year at over twenty-five stations at over fifty depths. The Hamilton Harbour Research project will have archived over a gigabyte of heterogeneous image data by the end of this year and the visualization tool permits a clear understanding of these data at a glance. The volumetric patterns of increasing temperature and decreasing oxygen content can be easily followed and understood. The relations between observed patterns and the complex circulations are also made evident through this approach.

Visualization showed three dimensional patterns not visible otherwise. Temperature data, for example, show the water masses present in the lake and their movement to the bay. The process of oxygen depletion can be followed in the three dimensional space and the number of voxels with given water quality properties can be easily computed for further analysis. A benefit of visualization is also that the data can be viewed interactively from different perspectives and with different viewing options thus increasing the interaction between scientist and the data.

5 ACKNOWLEDGMENTS

Murray Charlton provided his unpublished data collected in Hamilton Harbour. Jackie Dowell prepared the FORTRAN program to convert the data to the volume visualization format and create the appropriate masks. Robert Coker gave useful suggestions, helped with the editing and gave creative assistance in the rendering process. Spyglass corporation provided a beta version of their program Slicer.

6 REFERENCES

Adams, L., Gilsback, J.M., Krybus, W., Meyer-Ebrecht, D., Mosges, R. and Schlondorff, G., (1990) CAS-A navigation support for surgery. In K.H. Hoehne, H. Fuchs and S.M. Pzier (eds.) *3D Imaging in Medicine: Algorithms, Systems, Applications,* Springer Verlag, Berlin, 411-423.

Assel, R.A., Croley, T.E. II, and Schneider, K. (1994) Computer visualization of long-term average Great Lakes temperatures and ice cover. *J. Great Lakes Res.*, **20**, 771-782.

Bergeron, R.D. and Kaufman, A.E. (eds.) (1994) *Proceedings, Visualization '94.* IEEE Computer Society Press, Los Alamitos, CA.

Blinn, J.F. (1982) Light reflection functions for simulation of clouds and dusty surfaces. *Computer Graphics*, **16**, 21-29.

Charlton, M. (1994) Pers. Comm. National Water Research Institute, Canada Centre for Inland Waters, Burlington, Ontario, Canada.

Drebin, R.A., Carpenter, L., and Hanraham, P. (1988) Volume rendering, *Computer Graphics,* **22**, 64-75.

Foley, J.D., van Dam, A., Feiner, S.K. and Hughes, J.F. (1991). *Computer Graphics: Principles and Practice,* Addison-Wesley Publishing Co., Reading, Mass.

Foley, T.A., Hagen, H. and Nielson, G.M. (1993) Visualizing and modeling unstructured data. *The Visual Computer,* **9**, 439-449.

Frenkel, K.A. (1989) *Volume rendering.* Communications of the ACM, **32**, 426-435.

Greenleaf, J.F., Tu, T.S. and Wood, E.H. (1970). Computer-generated oscilloscopic images and associated techniques for display and study of the pulmonary blood flow, IEEE Trans. Nucl. Sci., NS-**17**, 353-359.

Halfon, E. and Howell, M. (1995) Visualization of limnological data as two- and three-dimensional computer generated animations. *J. Biol. Systems* , **4**, 3-25.

Halfon, E., Tartari, G., Howell, M., and Binelli, A. 1993. Visualization and limnological interpretation of liming treatments in a deep lake, Lake Orta, Italy, using computer generated animations. In IFIP, WG 5.11 Working conference on Computer Support for Environmental Assessment, CSEIA '93, villa Olmo, Como, Oct. 6-8, 1993. pp. 37-48.

Herman, G. T. and Udupa, J.K. (1983) Display of 3D digital images: computational foundations and medical applications. Proc. Medcomp '82, IEEE Computer Society Press, 308-314.

Hibbard, W and Santek, D. (1989) Visualizing large data sets in the Earth Sciences. IEEE Computer, **22**, 53-57.

ICLEI (1990) World Congress of Local Governments for a Sustainable Future, New York, 1990. International Council for Local Environmental Initiatives, World Secretariat, City Hall, East Tower, 8th Floor, Toronto, Ontario, Canada M5H 2N2.

IMSL (1987) Math/Library^T.M.^ User's manual. FORTRAN subroutines for mathematical applications. IMSL, 2500 ParkWest Tower One, 2500 CityWest Blvd., Houston, Texas 77042-3020, U.S.A.

Kaufman, A. (1991a) Introduction to Volume Visualization in A. Kaufmann (ed.) *Volume visualization,* IEEE Computer Society Press, Los Alamitos, California, pp. 1-18.

Kaufman, A. (1991b) *Volume visualization.* Los Alamitos, California: IEEE Computer Society Press, 477 pp.

Kaufman, A. 1992. *Introduction to volume visualization. ACM Siggraph '92, Course Notes, 285 pp.*

Levoy, M. 1988. Display of surfaces from volume data. IEEE Computer Graphics and Applications, **8**, 29-37.

Manley, T.O. and Tallet, J.A. 1990. Volumetric Visualization: an effective use of GIS technology in the field of oceanography. *Oceanography,* **3**, 23-29.

McCormick, B.H., DeFanti, T.A., and Brown, M.D. (1987) Visualization in scientific computing, *Computer Graphics,* **21** (special issue).

Mercurio, P.J. (1991) The data visualizer, *Pixel,* July/August, 31-35.

Rhodes, M.L., Kuo, Y., and Rothman, S.L.G. (1987) An application of computer graphics and networks to anatomic model and prosthesis manufacturing. *IEEE Computer Graphics,* **7**, 12-25.

Shirley, P and Neeman, H. (1989) Volume visualization at the center for supercomputing research and development. Proc. Chapel Hill Workshop on Volume Visualization, Univ. of North Carolina Press, Chapel Hill, N.C., pp. 17-20.

Spyglass (1994) Slice, Spyglass Inc. 1800 Woodfield Dr., Savoy, Illinois.

Upson, C and Keeler, M. (1988) V-BUFFER: Visible volume rendering, *Computer Graphics,* **22**, 59-64.

INDEX OF CONTRIBUTORS

KEYWORD INDEX